AF724942

SEEING THINGS

SEEING THINGS

The Philosophy of Reliable Observation

Robert Hudson

OXFORD
UNIVERSITY PRESS

OXFORD
UNIVERSITY PRESS

Oxford University Press is a department of the University of Oxford.
It furthers the University's objective of excellence in research, scholarship,
and education by publishing worldwide.

Oxford New York
Auckland Cape Town Dar es Salaam Hong Kong Karachi
Kuala Lumpur Madrid Melbourne Mexico City Nairobi
New Delhi Shanghai Taipei Toronto

With offices in
Argentina Austria Brazil Chile Czech Republic France Greece
Guatemala Hungary Italy Japan Poland Portugal Singapore
South Korea Switzerland Thailand Turkey Ukraine Vietnam

Oxford is a registered trademark of Oxford University Press
in the UK and certain other countries.

Published in the United States of America by
Oxford University Press
198 Madison Avenue, New York, NY 10016

© Oxford University Press 2014

All rights reserved. No part of this publication may be reproduced, stored in a
retrieval system, or transmitted, in any form or by any means, without the prior
permission in writing of Oxford University Press, or as expressly permitted by law,
by license, or under terms agreed with the appropriate reproduction rights organization.
Inquiries concerning reproduction outside the scope of the above should be sent to the
Rights Department, Oxford University Press, at the address above.

You must not circulate this work in any other form
and you must impose this same condition on any acquirer.

Library of Congress Cataloging-in-Publication Data
Hudson, Robert (Robert Glanville), 1960–
Seeing things : the philosophy of reliable observation / Robert Hudson.
pages cm
Includes bibliographical references and index.
ISBN 978–0–19–930328–1 (hardback : alk. paper) — ISBN 978–0–19–930329–8 (updf)
1. Observation (Scientific method) 2. Science—Philosophy. I. Title.
Q175.32.O27H83 2014
001.4'2—dc23
2013001191

1 3 5 7 9 8 6 4 2
Printed in the United States of America
on acid-free paper

In memory of Robert Butts, Graham Solomon, and Rob Clifton

CONTENTS

Preface *xi*
Introduction *xiii*

1. For and Against Robustness 1
The No-Miracles Argument for Robustness 2
Probabilistic Approaches to Robustness 8
Pragmatic Approaches to Robustness 25
Epistemic Independence Approaches to Robustness 36
Summary 51

2. The Mesosome: A Case of Mistaken Observation 52
Introducing the Mesosome: Rasmussen and Culp 55
The Mesosome Experiments 59
Reliable Process Reasoning 65
Rasmussen's Indeterminism 72

3. The WIMP: The Value of Model Independence 79
Dark Matter and WIMPs 81
DAMA's Model-Independent Approach 82
Model-Dependent Approaches to Detecting WIMPS 88
An Historical Argument Against Robustness 93
Reliable Process Reasoning 97

4. Perrin's Atoms and Molecules 103
Perrin's Table 104
The Viscosity of Gases 107
Brownian Movement: Vertical Distributions in Emulsions 116
Brownian Movement: Displacement, Rotation and Diffusion of Brownian Particles 124
Taking Stock 130
Perrin's Realism about Molecules 134

5. Dark Matter and Dark Energy 139
Dark Matter and the Bullet Cluster 142
Type Ia Supernovae and Dark Energy 150
Defeating Systematic Errors: The Smoking Gun 159
Robustness in the Dark Energy Case 166

6. Final Considerations Against Robustness 169
Independence and the Core Argument 170
The Need for Independence Does Not Equal the Need for Robustness 174
The Converse to Robustness Is Normally Resisted 179
The Corroborating Witness: Not a Case of Robustness 182

No Robustness Found in Mathematics and Logic 189
Robustness Fails to Ground Representational Accuracy 195
The Sociological Dimension of Robustness 198

7. Robustness and Scientific Realism 201
The No-Miracles Argument for Scientific Realism 202
In Support of Theoretical Preservationism 204
Objections to Theoretical Preservationism 208
Realism, the Pessimistic Meta-Induction and Preservationism 218
The Improved Standards Response: 'Methodological Preservationism' 226

Conclusion 243

Appendix 1 *249*
Appendix 2 *251*
Appendix 3 *253*
Appendix 4 *255*
Bibliography *259*
Index *267*

PREFACE

Some of the material in this book has been adapted from previously published work. The argument by cases early in chapter 1 and the bulk of chapter 3 draw from my paper 'The Methodological Strategy of Robustness in the Context of Experimental WIMP Research' (*Foundations of Physics*, vol. 39, 2009, pp. 174–193). The latter sections of chapter 1 on epistemic independence is a reworking my paper 'Evaluating Background Independence' (*Philosophical Writings*, no. 23, 2003, pp. 19–35). The first half of chapter 2 borrows heavily from my paper 'Mesosomes: A Study in the Nature of Experimental Reasoning' (*Philosophy of Science*, vol. 66, 1999, pp. 289–309), whose appendix is the basis of Appendix 4, and the second half of chapter 2 draws from 'Mesosomes and Scientific Methodology' (*History and Philosophy of the Life Sciences*, vol. 25, 2003, pp. 167–191). Finally, the first section of chapter 6 (Independence and the Core Argument) uses material from my 'Perceiving Empirical Objects Directly' (*Erkenntnis*, vol. 52, 2000, pp. 357–371).The rest of the material in the book has not previously been published.

My critique of Franklin and Howson (1984) in chapter 1 derives from a presentation of mine, 'An Experimentalist Revision to Bayesian Confirmation Theory,' at the 1993 Eastern Division meeting of the American Philosophical Association in Atlanta, Georgia. The commentator for that paper was Allan Franklin, and I am grateful both for his comments at that time and for subsequently inviting me to visit the University of Colorado in March 1994 as a Research Associate in the Department of Physics. In the spring of 1995 I presented the paper 'Notes Towards Representing the Uncertainty of Experimental Data in Bayesian Confirmation Theory' at the annual meeting of the Committee on the History and Philosophy of Science arranged by Allan and held at University of Colorado at Boulder. Though the material that formed the basis of that talk was never published, it inspired some debate among the participants there, notably Graham Oddie, Steve Leeds, and Clark Glymour. This debate prompted Graham to send around a detailed letter outlining a new way to introduce experimental uncertainty into Bayesian calculations (inspired, he notes, by comments made by Steve), and it is to this letter that I refer in chapter 1. I am grateful for the interest Graham, Steve, Clark, and Allan showed in my work at that time.

Throughout the many years before landing a permanent appointment at the University of Saskatchewan, I relied heavily on the support of many letter writers, especially William Harper, John Nicholas, and Murray Clarke. I wish to express my sincerest thanks to Bill, Nick, and Murray for their support during that time. I also wish to thank my colleagues at the Department of Philosophy at the University of Saskatchewan for a stimulating philosophical environment. This work was supported by a successive series of three Standard Research Grants obtained from the Social Sciences and Humanities Research Council of Canada, for which I am grateful. Additionally, detailed comments by readers from Oxford University Press proved extremely helpful. Finally, I thank my family for their love and support.

INTRODUCTION

You read in a local newspaper that alien life has been discovered, and you are suspicious about the accuracy of the report. How should you go about checking it? One approach might be to get another copy of the same newspaper and see if the same article appears. But what good would that be, if the copies come from the same printing press? A better alternative, many assert, would be to seek out a different news source, a different newspaper perhaps, and check the accuracy of the news report this way. By this means, one can be said to 'triangulate' on the story; by using multiple sources that confirm the story, one's evidence can be said to be 'robust'.

The current orthodoxy among philosophers of science is to view robustness as an effective strategy in assuring the accuracy of empirical data. A celebrated passage from Ian Hacking's (1983) *Representing and Intervening* illustrates the value of robustness:

> Two physical processes—electron transmission and fluorescent re-emission—are used to detect [dense bodies in red blood cells]. These processes have virtually nothing in common between them. They are essentially unrelated chunks of physics. It would be a preposterous coincidence if, time and again, two completely different physical processes produced identical visual configurations which were, however, artifacts of the physical processes rather than real structures in the cell. (201)

Here, identical visual configurations are produced through different physical processes—that is, they are produced 'robustly'—and Hacking's point is that there is a strong presumption in favour of the truth of robust results. The reason for this presumption is one's doubt that one would witness an identical observational artifact with differing physical processes. A similar viewpoint is expressed by Peter Kosso (1989), who comments:

> The benefits of [robustness] can be appreciated by considering our own human perceptual systems. We consider our different senses to be independent to some degree when we use one of them to check another. If I am uncertain whether what I see is a hallucination or real fire, it is less convincing of a test simply to look again than it is to hold out my hand and feel the heat. The independent account is the more reliable, because it is less likely that a systematic error will infect both systems than that one system will be flawed. (246)

Similar to Hacking's, Kosso's view is that, with robust results, the representational accuracy of the results best explains why they are retrieved with differing physical processes.

Of course, the value of this sort of argument depends on the relevant physical processes being 'different' or, more exactly, 'independent'. The question of what we mean here by 'independent' is a substantive one. We can start by emphasizing that our concern is, mainly, independent physical processes and not processes utilizing independent theoretical assumptions. To be sure, if different physical processes are being used to generate the same observational data, then it is very likely that the agents using these processes will be employing differing theoretical assumptions (so as to accommodate the differences in processes being used). It is possible that observers, by employing differing theoretical assumptions, thereby end up deploying different physical processes. But it is characteristic of scientific research that, when we talk about different observational procedures, we are ultimately talking about different physical processes that are being used to generate observations and not (just) different interpretations of an existing process. In this regard, we depart from the views of Kosso (1989), who sees the independence of interpretations of physical processes (and not the independence of the physical processes themselves) as more central to scientific objectivity. He says:

> The independence of sensory systems is a physical kind of independence, in the sense that events and conditions in one system have no causal influence on events and conditions in another. But the independence relevant to objectivity in science is an epistemic independence between theories. (246)

It follows on Kosso's view that the main threat to objectivity in science stems from the theory dependence of observation: He takes there to be value in generating identical observational results using differing theoretical assumptions—a requirement called 'epistemic independence'—to avoid a case in which a particular theory rigs the results of an observational procedure in its favour. Conversely, the classification I am mostly concerned with emphasizes the 'physical independence' of observational procedures (which might or might not be associated with the epistemic independence of the procedures). In this book we have the opportunity to criticize both kinds of robustness reasoning, one based on independent physical processes and the other based on independent interpretations (of physical processes).

The strategy of robustness reasoning envisioned by Hacking (1983) and Kosso (1989) can be succinctly expressed as follows: 'If observed result O is generated using independent observational processes, then there is strong evidence on behalf of the reliability of these processes, and so the truth of O has strong justification as well'. This strategy enjoys wide support in the philosophical literature and is periodically endorsed by scientists themselves in their more philosophical moments. Prominent philosophical advocates of robustness include Nancy Cartwright (1983) and Wesley Salmon (1984), each of whom cite famous work by the scientist Jean Perrin proving the existence of atoms as a paradigm example of how a scientist can, and should, use robustness reasoning. We examine below the arguments Perrin gives in 1910 and 1916 and find that his arguments are not in fact examples of robustness reasoning once we read them closely, even though Perrin, in reflecting on these arguments, views them this way himself. Similarly, one might be inclined to read John Locke as a supporter of robustness reasoning if one is not a careful student of a certain passage in John Locke's *Essay Concerning Human Understanding* (Book 4, chapter 11, section 7), a passage that evidently influenced Kosso's thinking on the topic. In that passage Locke (1690) says:

> Our senses in many cases bear witness to the truth of each other's report, concerning the existence of sensible things without us. He that sees a fire, may, if he doubt whether it be anything more than a bare fancy, feel it too; and be convinced, by putting his hand in it. (330–331; italics removed)

This is once more Kosso's fire example referenced above. But notice what Locke (1690) continues to say when he explains the benefit of an alternate source of evidence:

> [In feeling fire, one] certainly could never be put into such exquisite pain by a bare idea or phantom, unless that the pain be a fancy too: which yet he cannot, when the burn is well, by raising the idea of it, bring upon himself again. (331; italics removed)

In other words, it is not simply the convergence of the testimonies of sight and touch that speak on behalf of there really being a fire there but rather the fact that putting one's hand in a fire is a far better, more reliable test for the reality of a fire than visual observation—the latter, but not the former, can be fooled by 'a bare idea or phantom'. So, for Locke, the value in utilizing an alternate observational strategy does not derive from some special merit of having chosen an observational procedure that is simply independent and nothing more than that. The value of multiplying observational procedures depends on the character of the independent procedures themselves, on whether they already have an established reliability that can address potential weaknesses in the procedures already being deployed. The main task of this book could be thought of as a development of this Lockean perspective.

In setting forth this critique of robustness, my first step is to examine why philosophers (and others) are inclined to believe in the value of robustness. To this end I examine in chapter 1 a variety of philosophical arguments in defence of robustness reasoning. A number of these arguments are probabilistic; some arguments, mainly due to William Wimsatt (1981), are pragmatic; others follow Kosso's (1989) epistemic definition of independence. Although I conclude that all these approaches are unsuccessful, there is nevertheless a straightforward argument on behalf of robustness that is quite intuitive. I call this argument the 'core argument' for robustness, and the full refutation of this argument occurs in chapter 6.

As I do not believe that my anti-robustness arguments can be carried on exclusively on philosophical, a priori grounds, the full critique of robustness and the beginnings of a better understanding of how scientists justify the reliability of observational data must engage real scientific episodes. To this end I spend chapters 2 through 5 looking at five different scientific cases. The first case, discussed in chapter 2, deals with the mistaken discovery of a bacterial organelle called the mesosome. When electron microscopes were first utilized in the early 1950s, microbiologists found evidence that bacteria, previously thought to be organelle-less, actually contained midsized, organelle-like bodies; such bodies had previously been invisible with light microscopes but were now appearing in electron micrographs. For the next 25 years or so, the structure, function and biochemical composition of mesosomes were active topics of scientific inquiry. Then, by the early 1980s it came to be realized that mesosomes were not really organelles but were artifacts of the processes needed to prepare bacteria for electron-microscopic investigation. In the 1990s, philosopher Sylvia Culp (1994) argued that the reasoning microbiologists ultimately used to demonstrate the artifactual nature of mesosomes was robustness reasoning. In examining this case, I argue that robustness reasoning wasn't used by microbiologists to show that mesosomes are artifacts. (In fact, if microbiologists had used robustness, they would have likely arrived at the wrong conclusion that mesosomes are indeed real.) Alternatively, in examining the reasoning of microbiologists, I see them arguing for the artifactual nature of mesosomes in a different way, using what I term 'reliable process reasoning'.

In chapter 3 I consider a different case study, this time involving the search for the particle that is believed to constitute cosmological dark matter, called the WIMP (weakly interacting dark matter). Various international research teams are currently engaged in the process of searching for WIMPs, with the majority of teams arriving at a consensus that WIMPs have not (yet) been detected. On that basis there is room to argue robustly for the claim that WIMPs don't exist, as the no-detection result has been independently arrived at by a number of researchers. However, as we shall see, such a form of robustness reasoning does not impel the thinking of these teams of astroparticle physicists. Meanwhile, there is unique a group of astroparticle physicists who claim to have observed WIMPs using what they call a model-independent approach, an approach they believe to be more reliable than the model-dependent approaches employed by the many groups who have failed to observe WIMPs. I believe the significance of this model-independent approach is best understood as illustrating a form of reliable process reasoning as this notion is set forth in chapter 2. Robustness reasoning, by comparison, has little relevance to this case despite the fact that it has obvious application.

Chapter 4 deals what is often thought to be a classic instance of a scientist using robustness reasoning—Jean Perrin's extended argument for the reality of atoms (and molecules). Perrin lists a number of different methods for calculating Avogadro's number, and as they all converge within an acceptable degree of error, Perrin asserts that he has found a rigorous basis for inferring that atoms exist. Perrin even describes his reasoning in a way strongly reminiscent of robustness when introducing and summarizing his arguments. However, once we look closely at his reasoning in both *Brownian Movement and Molecular Reality* (Perrin 1910) and *Atoms* (Perrin 1916 [4th edition] and Perrin 1923 [11th edition]), reasoning that purports to establish on empirical grounds the atomic theory of matter, we find that robustness is not used by Perrin after all. Consequently, it turns out that one of the pivotal historical case studies in support of robustness reasoning is undermined, despite the many assured allusions to this case by such pro-robustness supporters as Ian Hacking (1983), Nancy Cartwright (1983), Wesley Salmon (1984), Peter Kosso (1989) and Jacob Stegenga (2009). As I argue, Perrin is engaged in a different form of reasoning that I call 'calibration', which could be mistaken for robustness reasoning if one isn't cautious in how one reads Perrin. Calibration, I argue, plays a key role in Perrin's realism about atoms and molecules.

The final two cases are discussed in chapter 5. Here I return to the science of dark matter, but now at a more general level, and consider arguments raised on behalf of the reality of dark matter, leaving to one side the question of the composition of dark matter (assuming it exists). Once again, obvious robustness arguments are bypassed by astrophysicists who alternatively focus on a different reasoning strategy that I call 'targeted testing'. Targeted testing comes to the forefront when we consider one of the pivotal pieces of evidence in support of dark matter, evidence deriving from the recent discovery of the cosmological phenomenon called the Bullet Custer. Targeted testing is also utilized in the second case study discussed in chapter 5 dealing with the recent (Nobel Prize–winning) discovery of the accelerative expansion of the universe, an expansion said to be caused by a mysterious repulsive force called dark energy. The dark energy case is interesting due to the fact that a prominent participant of one of the groups that made this discovery, Robert Kirshner, argues explicitly and forcefully that robustness reasoning (in so many words) was fundamental to justifying the discovery. Similar to what we find with Perrin, my assessment is that Kirshner (2004) misrepresents the reasoning underlying the justification of dark energy, an assessment at which I arrive after looking closely at the key research papers of the two research groups that provide observational evidence for the universe's accelerative expansion. I argue that astrophysicists use, similar to what occurred in the Bullet Cluster case, a form of targeted testing—and do so to the neglect of any form of robustness reasoning.

With our discussion of real cases in science behind us, chapter 6 picks up again the argument against robustness begun in chapter 1 and provides a series of arguments against robustness that are in many respects motivated by our case studies. To begin, the core argument for robustness that was deferred from chapter 1 is reintroduced and found to be questionable due to our inability to adequately explain what it means for two observational processes to be independent of one another in a way that is informative. There are, I contend, identifiable benefits to independent lines of empirical inquiry, but they are benefits unrelated to robustness (such as the motivational benefits in meeting empirical challenges on one's own, independently of others). Moreover, I express concern in this chapter that supporters of robustness reasoning say precious little about the details of how this reasoning is to be applied. For example, which of the many possible independent procedures should be utilized, or doesn't this matter? How different should these alternate procedures be, and how many of them should be used—or is this number open-ended? In the literature, robustness reasoning is often presented in such an abstract form that how to use it effectively in practical terms is left unclear. For example, guidance is seldom given on how we should represent a robust, observed result. Even granting the existence of a common element of reality that independently causes through different procedures the same observed result, such a convergence isn't informative to us without an accurate description of this common element, yet the details of this description inevitably lead us beyond the purview of what robustness has the capacity to tell us. To close chapter 6, and in recognition of the fact that robustness reasoning is highly esteemed by many philosophers and the occasional scientist, I suggest some sociological reasons that account for its evident popularity.

With my critique of robustness completed by chapter 6, my next step in chapter 7 is to apply my negative assessment of robustness to some recent moves that have been made in the (scientific) realism/antirealism debate. After setting forth familiar reasons for an antirealist view of science, I recount a popular defense of realism based on the doctrine of 'preservationism', often instantiated as a form of 'structuralism'. Both preservationism and structuralism, I argue, are flawed because the legitimacy of each is based on grand form of historical, robustness reasoning. Over the course of history, it is said, many scientific theories rise to prominence and then fade away, leading the antirealist to conclude that no one theory is a legitimate candidate for a realist interpretation. In response to this pessimistic view, the preservationist (and structuralist) suggests that there are certain components of these (transiently) successful scientific theories that are retained (perpetually, in the best case) within future, successful scientific theories. With structuralism, more precisely, the claim is that these preserved components are structural, where the meaning of 'structure' is variously interpreted (such variations having no bearing on my argument). It is then about such preserved elements that preservationists (and structuralists) claim we are in a position to be realist. As it were, each successful, though transient scientific theory is just one method of displaying the reality of these preserved elements, and the fact that a number of transient, successful theories contain these preserved elements indicates that these elements represent some aspect of reality. Why else, one might ask, do they keep showing up in a progression of successful theories? Reasoning in this way has a clear affinity to the form of robustness reasoning we described with regard to observational procedures: The differing theories are analogous to independent observational procedures, and the preserved elements correspond to the unique observed results that emanate from these procedures. The accuracy of this analogy is justified once we consider the sorts of critiques that have been launched against preservationism, such as by the philosophers Hasok Chang (2003) and Kyle Stanford (2003, 2006), who raise doubts about the independence of the theories containing preserved elements. Briefly, my claim is that, if the analogy between preservationism and observational robustness holds up, then the arguments I have adduced against robustness apply analogously to preservationism (and to structuralism), which means that these ways of defending scientific realism are undermined.

If we lose the authority of preservationism (and correlatively structuralism) as a response to antirealism, we need new grounds on which to defend scientific realism. The remainder of chapter 7 is devoted to the task of proposing and defending just such new grounds. My new version of scientific realism I label 'methodological preservationism'. It is a realism that is inspired by the recent writings of Gerald Doppelt (2007). It is also a realism that is heavily informed by the case studies that form the core of this book. The resultant realism is characterized by a form of cumulativism, though one very much different from the form of preservationism I describe above. According to the cumulativism I defend, what are preserved over time are not privileged scientific objects but privileged observational methods. There are, I argue, certain observational methods whose reliability, understood in a general sense, is largely unquestioned and that we can anticipate will remain unquestioned into the future. These methods serve as observational standards that all subsequent theorizing must respect, wherever such theorizing generates results that are impacted by the outputs of these methods. The primordial such standard is naked-eye (i.e., unenhanced) observation. This is an observational procedure whose reliability (in general terms) is unquestioned and whose reliability will continue to be unquestioned as long as humans remain the sort of animals they currently are (e.g., if in the future we don't evolve different forms of 'naked' observational capacities that reveal a very different world). The point of being a preserved methodology is that it is assumed to provide a reliable picture of the world, and thus there is a prima facie assumption in favour of the reality of whatever it is that this methodology portrays. For example, with naked-eye observation, there is a prima facie assumption in favour of the reality of the macroscopic, quotidian world, containing such things as trees, chairs, tables and the like. Still, the scientific consensus about what naked-eye observation reveals is changeable and has occasionally changed in the past; what counts as real according to naked-eye observation is not fixed in time, since views about the components of the macroscopic world can vary. To take an obvious example, early mariners upon seeing a whale likely considered it to be a (big) fish; our view now is that whales are in fact mammals. Nevertheless, for the most part the taxonomy of the macroscopic world has been fairly constant, though not because the objects in this world occupy a special ontological category. Rather this ontological stability is a byproduct of the stable, established credentials of the process by which we learn about these things—naked-eye observation. It is a process whose authority has been preserved over time, and though what it reveals has been fairly constant as well, there is no necessity that this be true. What I show in this chapter is that the sort of methodological authority ascribed to naked-eye observation is extendable to forms of mediated observation. For instance, both telescopy and microscopy are regarded as possessing an inherent reliability: In researching the structure of physical matter, it is granted by all that looking at matter on a small scale is informative, just as we all agree that using telescopes is a valuable method for investigating distant objects. In my view, we find in science a progression of such authoritative observational technologies, starting from the base case, naked-eye observation, and incorporating over time an increasing number of technological and reason-based enhancements whose merits have become entrenched and whose usefulness for future research is assured.

Before proceeding with our investigation let me make two small, clarificatory points. First, we should be clear that the term 'robustness' in the philosophy of science literature carries different, though related meanings, all connected by the fact that each 'describes a situation where one thing remains stable despite changes to something else that, in principle, could affect it' (Calcott 2011, 284). In this book we mean 'robustness' strictly in what Calcott (2011) calls the 'robust detection' sense, where

> a claim about the world is robust when there are multiple, independent ways it can be detected or verified.... For example, different sensory modalities may deliver consistent information about the world, or different experimental procedures may produce the same results. (284)

Woodward (2006) calls this sense of robustness 'measurement robustness', and argues for 'the undoubted normative appeal of measurement robustness as an inductive warrant for accepting claims about measurement', using as an explanation for this normative appeal an argument that is very much like, if not identical to what I call the 'core argument' for robustness (234). In contrast, one can also mean robustness in the 'robust theorem' (Calcott) or 'inferential robustness' (Woodward) sense. This is the sense one finds in Levins (1966), which has been subsequently critiqued by Orzack and Sober (1993) and by Woodward (2006). As Calcott (2011) explains, in this sense,

> a robust theorem is one whose derivation can be supported in multiple ways,... mostly discussed in the context of modelling and robustness analysis. To model a complex world, we often construct models—idealised representations of the features of the world we want to study.... [Robustness] analysis identifies, if possible, a common structure in all the models, one that consistently produces some static or dynamic property. (283)

Woodward expresses the concern that the merits of measurement robustness do not carry over to inferential robustness (2006, 234), and cites Cartwright (1991) as a source for these concerns (2006, 239, footnote 13). But for all their consternation about inferential robustness, neither Woodward nor Cartwright express any qualms about the epistemic value of measurement robustness, and each cite Perrin as a classic illustration of this form of reasoning (Woodward 2006, 234; Cartwright 1991, 149–150, 153). Ironically, I believe some of the concerns harboured by Woodward and Cartwright regarding inferential robustness carry over to measurement robustness, which motivates me to return to the issue of inferential robustness at two places: first, in chapter 1 in my discussion of a Wimsattian, pragmatic approach to defending (measurement) robustness, and secondly, in chapter 6 where I examine the potential for robustness arguments in mathematics and logic. Finally, for the remainder of the senses of 'robustness' on offer (for example, Woodward 2006 cites in addition 'derivational' and 'causal' notions of robustness, where the latter is likely what Calcott 2011 means by 'robust phenomena'), we leave discussion of them aside.

The second, clarificatory point I wish to make is that throughout this book I often refer to 'observational' processes and procedures, and omit reference to the 'experimental'. This is because, to my mind, there is no difference in kind between observational and experimental processes—the former term is a generalization of the latter, where the latter involves a more dedicated manipulation of a physical environment to allow new or innovative observations to be made. Here I differ from some who regard observation as 'passive' and experimentation as 'active', and so as fundamentally different. My view is that once an experimental mechanism is set up, the results are 'passive' observations just as with non-experimental setups (an experimenter will passively see a cell under a microscope just as we now passively see chairs and tables). Moreover, even with naked-eye observation, there is at the neurophysiological level an enormous amount of active manipulation of the data, and at the conscious and sub-conscious levels a great deal of cognitive manipulation as well. So I find no fundamental difference between enhanced ('experimental') and unenhanced ('naked-eye') observing, and opt wherever convenient to use the more general term 'observational'.

Chapter 1

For and Against Robustness

Over the years, robustness reasoning has been supported by many philosophers (and philosophically minded scientists), and there has been various attempts to put the legitimacy of robustness reasoning on firm footing (though for many the legitimacy of robustness is an obvious truth that need not be argued for). Have these attempts been successful? This is the question we address in this chapter, and unfortunately for robustness theorists my response is in the negative—each of the strategies we examine that strive to put robustness reasoning on firm footing suffers important flaws. But my task in this book is not entirely negative. Later on in the book, after examining a number of historical case studies, I suggest some methods that scientists actually use to ensure the accuracy of observational data, methods that can (deceptively) appear to involve robustness reasoning. In other words, the reader will not be abandoned withouta story about how scientists go about ensuring the accuracy of observational data.

Our immediate task, nevertheless, is to gain a grasp on various arguments that have been given for the cogency of robustness reasoning. In the Introduction we saw the outline of an argument (due to Ian Hacking and Peter Kosso) for the value of robust, observational results: Where different physical processes lead to the same observed result, the representational accuracy of this result seems to be the best (or even only) explanation of this convergence. I call this the 'no-miracles' argument for robustness, and in the next section I offer an abstract (and by no means conclusive) argument against this approach. In subsequent sections I look at three alternative, different approaches to justifying robustness—approaches that are (a) probabilistic, (b) pragmatic and (c) based on epistemic independence. The probabilistic approaches we examine utilize the resources of (typically Bayesian) probability theory to show that robust observations have a greater likelihood of being true. Pragmatic approaches focus on the ability of robust results to resist refutation (leaving aside the related question of whether such resistance is a sign of truth). Finally, epistemic independence approaches find robustness reasoning to be an antidote to the theoretical circularity that, for some, can undermine the objectivity of empirical testing. All these approaches, I argue, have their irremediable weaknesses. Still, there is a fundamental philosophical insight underlying robustness reasoning that many have found compelling, an insight encapsulated in what I call the 'core' argument for robustness. I deal directly with the core argument in chapter 6, after examining of a number of historical case studies in chapters 2 through 5.

THE NO-MIRACLES ARGUMENT FOR ROBUSTNESS

When different observational processes lead to the same observed result, the no-miracles argument for robustness leads to the conclusion that the observed result is (likely) factually true if, given the description of the situation, it is highly unlikely that such convergence would happen by accident (such as if the result were an artifact of each of the observational processes). This argument has clear affinity to the popular argument for scientific realism by the same name, according to which the best explanation for the success of science over time is the (approximate) representational accuracy of science. One difference with the observational 'robustness' version of the argument is that, since it applies strictly to observational results, the relevant no-miracles argument has a narrower scope—that is, the relevant notion of success refers solely to the retrieval of convergent observational results, not to what could count as scientific success in general terms. There is the potential, then, for a more direct assessment of the quality of an observational, no-miracles robustness argument, with its narrower conception of empirical success.

I have attributed this observational, no-miracles robustness argument to Ian Hacking in light of the passage quoted in the Introduction, and here one might resist such an attribution on the grounds that Hacking (1983) in the same book explicitly disavows the epistemic force of an analogous, convergence no-miracles argument for scientific realism based on the ability of a theory to explain multiple, independent phenomena. Hacking cites as an instance of this 'cosmic accident argument' (as he calls it) the convergence since 1815 of various computations of Avogadro's number. This convergence (to a value of $60.23 \cdot 10^{22}$ molecules per gram-mole—see Hacking 1983, 54–55) is taken by many to constitute sufficient grounds for the accuracy of this computation and from here to the conclusion that molecules are real. Indeed, in chapter 4, we look at a version of this robustness argument attributable to Jean Perrin. For his part, Hacking is unimpressed with the realist conclusion drawn here, since he doesn't believe there are good grounds to say anything more than that the molecular hypothesis is empirically adequate, given the cited convergence—his view is that asserting the reality of molecules here simply begs the question on behalf of realism. He even questions whether 'is real' is a legitimate property, citing Kant's contention that 'existence is a merely logical predicate that adds nothing to the subject' (54). Given these views, what justification do we have for describing Hacking as an advocate of an observational no-miracles, robustness argument?

Such an interpretive question is resolved once we recognize that the sort of argument Hacking (1983) believes is portrayed in his 'red blood cell' example is not a cosmic accident argument at all but something different—what he calls an 'argument from coincidence'. According to this argument, dense bodies in red blood cells must be real since they are observed by independent physical processes, not because their postulation is explanatory of diverse phenomena. Indeed, he suggests that

> no one actually produces this 'argument from coincidence' in real life: one simply looks at the two (or preferably more) sets of micrographs from different physical systems, and sees that the dense bodies occur in exactly the same place in each pair of micrographs. That settles the matter in a moment. (201)

That is, for Hacking, the legitimacy of an argument from coincidence is so obvious (both to him and, presumably, to scientists generally) that one doesn't even need to state it. Nevertheless, he is aware of the striking similarity this argument has to the cosmic accident argument described above. So should Hacking's skepticism about the value of the latter sort of argument affect his attitude regarding the former argument from coincidence? He argues that the superficial similarity of these arguments should not conceal their inherent differences. First and foremost, these arguments differ as regards the theoretical richness of their inferred objects. With robust, observed results (i.e., the argument from coincidence), the inferred entity may be no more than that—an 'entity'. For example, the dense bodies in red blood cells as independently revealed through electron transmission microscopy and fluorescence microscopy Hacking understands in a highly diluted fashion. As he suggests, ' "dense body" means nothing else than something dense, that is, something that shows up under the electron microscope without any staining or other preparation' (1983, 202). As a result, these inferred entities play no substantive role in theoretically explaining observations of red blood cells. Hacking clarifies:

> We are not concerned with explanation. We see the same constellations of dots whether we use an electron microscope or fluorescent staining, and it is no 'explanation' of this to say that some definite kind of thing (whose nature is as yet unknown) is responsible for the persistent arrangement of dots. (202)

By comparison, with the cosmic accident argument, an elaborately understood theoretical entity is postulated, one that can richly explain observational data. For this reason Hacking asserts that we should not conflate the experimental argument from coincidence with the theoretical cosmic accident argument: Whereas the latter entertains detail that can render the argument dubious, the former, because it is theoretically noncommittal, has a greater assurance of truth.

Still we should be clear that the difference between the two forms of argument is a difference of degree, not a difference in kind. We can, if we like, describe robustness reasoning as a form of inference to the best explanation—for Hacking it is simply a theoretically uninformative inference, if we accept his view about the thin, theoretical character of experimentally discerned entities. It is moreover arguable that, for Hacking, the uniformativeness of the inference is related to his assumption of the trivially obvious, epistemic value of robust, experimental results (again, as he suggests, one hardly needs to 'produce the argument'). Closer examination of Hacking (1983) reveals in part why he is prone to trivialize robustness. It is because he works under the assumption that certain experimental approaches can independently be regarded (that is, independently of robustness considerations) as inherently reliable or unreliable. For instance, with respect to the dense bodies in red blood cells as revealed by electron microscopy, and considering the problem whether these bodies are 'simply... artifacts of the electron microscope', Hacking makes note of the fact that 'the low resolution electron microscope is about the same power as a high resolution light microscope', which means that, therefore, 'the [artifact] problem is fairly readily resolved' (200). Nevertheless, he notes, 'The dense bodies do not show up under every technique, but are revealed by fluorescent staining and subsequent observation by the fluorescent microscope' (200). That is, it is not (simply) the independence of two observational routes that is the key to robustness (presumably some of the techniques under which dense bodies fail to appear are independent of electron microscopy, in that they involve 'unrelated chunks of physics'). Instead it is for Hacking the prima facie assurance we have to begin with that a particular observational route is, to at least a minimal degree, reliable as regards a certain object of observation. In describing some of the experimental strategies used in comparing the results of electron transmission and fluorescent re-emission, he surprisingly comments that '[electron-microscopic] specimens with particularly striking configurations of dense bodies are... prepared for fluorescent microscopy' (201). Now, if the nonartifactuality of these dense bodies were a genuine concern, and if the plan was to use robustness reasoning to settle the question of artifactualness, the preparation of specimens with 'striking configurations of dense bodies' would be a puzzling activity. Where such bodies are artifacts, one would be creating specimens with a maximum degree of unreliability. So it must be Hacking's view that electron microscopy possesses a minimal level of reliability that assures us of the prima facie reality of dense bodies and that fluorescence microscopy is used to further authenticate the reliability of electron microscopy (as opposed to initially establishing this reliability).

The recognition that robustness reasoning assumes the (at least minimal) reliability of alternate observational routes and that it is ineffective at establishing this reliability to begin with forms a key part of my critique of robustness. For now, however, our goal is to assess the observational, no-miracles robustness argument, and I submit that the following argument exposes a key weakness with this argument. The argument proceeds by cases. We start by considering a situation where we have two different physical observational processes that converge on the same observed result. Each of these processes is either reliable or not, in (at least) the sense that each tends to produce a representationally accurate result, or it does not. So take the case where either both processes or at least one of them is unreliable. Then we are in no position to explain convergent observed results by reference to the representational accuracy of the processes since at least one of these processes tends not to generate representationally accurate results. In effect, if it so happens that both processes are generating the right results, this is indeed miraculous, considering at least one of the processes is unreliable. Accordingly, the miraculousness of the situation is not a feature that would need explaining away. So suppose, alternatively, that both processes are reliable. Then for each process there is a ready explanation for why it generates the relevant observed result—each process, being reliable, functions to produce representationally accurate results, and since the processes are being used to the same end, they produce the same observed results. Now, when we are confronted by this convergence of observed results using these processes, what should our conclusion be? Does this convergence need any special explaining? And in explaining this convergence, do we gain special support for the reliability of the processes and for the representational accuracy of the observed results? One might conjecture that this convergence is epistemically irrelevant since the reliability of the relevant processes is already assured. To illustrate this point, suppose we have a research group that produces observational data bearing on some theoretical claim and that this group is assured of the reliability of the process that produces this data and hence of the representational accuracy of the generated data. In such a case, would it matter to this group, as regards the reliability of the data, that there is another group of researchers that produces the same data using an entirely different physical process? Why would the first group be interested, epistemically speaking, in the work of other researchers generating the same result, given that for them the reliability of their work is already assured and they've already generated an accurate observed result?

At this point one might draw the inference that the observational, no-miracles argument for the value of robustness is ineffective. However, one could respond to this inference in the following way. Of course, if one knew that one's observational process was reliable, then (arguably) there would be no need to advert to another observational process in defending the reliability of the first process, even if we were aware of the reliability of this other process. But that's just the point: Because in many cases we lack knowledge of the reliability (or unreliability) of an observational process, we need an independent observational perspective to check on this process. By then noting that a new independent, observational process converges on the same observed result as the original process, we are in a position to cite the representational accuracy of this result along with the reliability of the two processes as a way of explaining this convergence.

This revised interpretation of the observational, no-miracles argument for robustness is important enough that I propose to call it the 'core argument' for robustness. It is an argument that will reappear as we explore various approaches that have been adduced to support robustness forms of reasoning, and a full refutation of this argument is presented in chapter 6, after we've had the chance in the intervening chapters to examine various historical case studies. For now, to give the reader an inkling of why I resist the core argument, consider a case where we lack a justified opinion regarding the reliability of each of two observational processes, a case where for all we know, both observational processes might be telling the truth, or only one might be, or neither of them is—we're simply unsure about which is the case. Given this situation, would it be appropriate where the two observational processes converge on the same convergent result to increase our confidence in the accuracy of the result? To me, this sounds like an uncertain way of proceeding, and it is unclear what we could learn from this situation. From a position of ignorance we would be drawing the conclusion that an observed result is more likely to be true given that it issues from multiple physical processes. Yet should we learn more—say, that one of the processes is more reliable than the other—it would then follow that this convergence is less significant to us (even if we assume the independence of the processes) for the simple fact that we naturally become more reliant on the testimony of the more reliable process. Similarly, if we learn that one of the processes is irrelevant to the issue of what is being observed, we would be inclined to outright dismiss the epistemic significance of the convergence. Overall it seems that it would be more advisable for an observer, when faced with uncertainty regarding the processes of observation, to work on improving her knowledge of these processes with an eye to improving their reliability rather than resting content with her ignorance and arguing instead on the basis of the robustness of the results.

It is for these kinds of reasons that I am suspicious of the value of the core argument for robustness. Further development of these reasons will occur later. In advance of examining these reasons, let us look at three other strategies for defending the value of robustness reasoning. The first approach is probabilistic, typically utilizing Bayesian confirmation theory, though I describe a likelihoodist approach as well. Although I argue that all of these probabilistic strategies are unsuccessful, they nevertheless provide interesting philosophical insights into the process of testing theories on the basis of observations.

PROBABILISTIC APPROACHES TO ROBUSTNESS

Our survey of different approaches to defending robustness begins with probabilistic strategies. One of the earliest and most effective probabilistic defenses of robustness can be found in Franklin and Howson (1984), whereas a very succinct version of this argument can be found in Howson and Urbach (2006, 126). Franklin and Howson reason on Bayesian grounds as follows.

We let E and E' be two different physical observational procedures (e.g., experiments) that individually generate the following two series of observed results: $e_1, e_2, e_3, \ldots e_m$ and $e_1', e_2', e_3', \ldots e_n'$ (the e_i and e_j' stand for the same result produced at subsequent times). We also assume that the likelihoods for each of these observed results given theoretical hypothesis h is unity (i.e., h entails all the e_i and e_j'), that is,

$$P(e_i/h) = P(e_j'/h) = 1$$

Franklin and Howson then formalize the notion of two observational procedures being different by means of two conditions: For some value of m,

$$P(e_{m+1}/e_1 \,\&\, e_2 \,\&\, e_3 \,\&\ldots\&\, e_m) > P(e'_j/e_1 \,\&\, e_2 \,\&\, e_3 \,\&\ldots\&\, e_m),$$

and for some value of n,

$$P(e_{n+1}'/e_1' \,\&\, e_2' \,\&\, e_3' \,\&\ldots\&\, e'_n) > P(e_i/e_1' \,\&\, e_2' \,\&\, e_3' \,\&..\&\, e'_n).$$

What these conditions are telling us is that, for observational procedures E and E', with continued repetitions yielding confirmatory results from one of these procedures, one comes to expect further such confirmatory results from this procedure, and thus at some point one has comparatively less expectation of a (confirmatory) observed result from the alternate procedure. A straightforward application of Bayes' theorem then yields the result:

$$\frac{P(h/e_1 \,\&\, e_2 \,\&\, e_3 \,\&\ldots\&\, e'_j)}{P(h/e_1 \,\&\, e_2 \,\&\, e_3 \,\&\ldots\&\, e_i)} = \frac{P(e_i/e_1 \,\&\, e_2 \,\&\, e_3 \ldots\&\, e_m)}{P(e'_j/e_1 \,\&\, e_2 \,\&\, e_3 \ldots\&\, e_m)} \qquad \text{(1a)}$$

(See Appendix 1 for proof.) Hence, at the point where continued repetitions of a confirmatory result from an observational procedure lead us to have comparatively less expectation of a (confirmatory) observed result from the alternate procedure—that is, $P(e_i/e_1 \,\&\, e_2 \,\&\, e_3 \,\&\ldots\&\, e_m) > P(e'_j/e_1 \,\&\, e_2 \,\&\, e_3 \,\&\ldots\&\, e_m)$—it follows (by the Bayesian positive relevance criterion) that h is better confirmed (that is, its posterior probability is increased more) by testing h with the observed result generated by the alternate procedure. In other words, evidence for h generated by E eventually becomes 'old' or 'expected,' and to restore a substantive amount of confirmation, new and unanticipated evidence is needed deriving from an independent observational procedure E'.

This is an elegant defense of the value of robust observational support for a hypothesis. However, it contains an oversight that is common to discussions of robustness and to philosophic discussions of the bearing of observed results on theories generally. The oversight is that when speaking of observed evidence for a hypothesis, one needs to consider whether the observational process generating this evidence is reliable and to what degree. Given such a consideration, Franklin and Howson (1984) need to factor in the comparative reliability of competing observational procedures when arguing for the claim that at some point in the collection of evidence one should switch observational procedures. For example, referring again to observational procedures E and E', if E' turns out to be a highly unreliable process, whereas E is highly reliable, then intuitively there is not much merit in switching procedures—a fact that Franklin and Howson's formalism fails to capture. How then might we incorporate this factor into their formalism? There are a number of ways by which one might do this, which we now explore.

To start, let's define a perfectly reliable experiment as one that generates the result e_i if and only if e_i is true. It then follows that where hypothesis h entails e_i, $P(e_i/h) = 1$. Now suppose that experiment E referred to above is less than perfectly reliable but more reliable than E'. We can formalize this difference as follows:

$$1 > P(e_i/h) > P(e_j'/h) > 0$$

That is, E is not perfect at tracking the truth of h but is better at it than E'. Now we ask the following question: If we are in the process of generating observed results using E, when is it better to switch from E to E'? That is, when is h better confirmed by evidence drawn from E' than from E? On the Bayesian positive relevance criterion, looking at a single application of each of E and E' and dropping subscripts for simplicity, e better confirms h than e', that is, $P(h/e) > P(h/e')$, if and only if

$$P(e/h)/P(e/-h) > P(e'/h)/P(e'/-h) \tag{1b}$$

(where $-h$ denotes the falsity of h; see Appendix 2 for proof). Assuming for simplicity that $P(e/-h) = P(e'/-h)$ (that is, E and E' are equally reliable at discerning e or e', respectively, where h is not true), it follows from a single application of each of these two experiments that evidence from a more reliable experiment better confirms a hypothesis than evidence from a less reliable experiment.

Now suppose we have repeated applications of E, leading to the results $e_1, e_2, e_3, \ldots e_m$. We saw that with a single application of E and E', e betters confirms h than e'. The question is, with repeated applications of E, when should we abandon E and look instead to E' to (better) confirm h? On the Bayesian positive relevance criterion, with repeated applications, $P(h/e_1 \,\&\, e_2 \,\&\, e_3 \,\&\ldots\&\, e_{m+1}) > P(h/e_1 \,\&\, e_2 \,\&\, e_3 \,\&\ldots\&\, e'_j)$ (i.e., e_{m+1} better confirms h than e'_j, after having witnessed a series of results $e_1, e_2, e_3, \ldots e_m$) if and only if

$$\frac{P(e_{m+1}/h \,\&\, e_1 \,\&\, e_2 \,\&\, e_3 \,\&\ldots\&\, e_m)}{P(e_{m+1}/-h \,\&\, e_1 \,\&\, e_2 \,\&\, e_3 \,\&\ldots\&\, e_m)} > \frac{P(e'_j/h \,\&\, e_1 \,\&\, e_2 \,\&\, e_3 \,\&\ldots\&\, e_m)}{P(e'_j/-h \,\&\, e_1 \,\&\, e_2 \,\&\, e_3 \,\&\ldots\&\, e_m)} \quad (1c)$$

(see Appendix 2 for proof). There are various ways one might interpret (1c), dependent on how one views the independence between E and E'. It may be that one views the outcomes of E as entirely probabilistically independent of the outcomes of E'. If so, $P(e'_j/h \,\&\, e_1 \,\&\, e_2 \,\&\, e_3 \,\&\ldots\&\, e_m) = P(e'_j/h) = P(e'/h)$, and similarly, $P(e'_j/-h \,\&\, e_1 \,\&\, e_2 \,\&\, e_3 \,\&\ldots\&\, e_m) = P(e'_j/-h) = P(e'/-h)$. Suppose, then, that $P(e'/-h) > P(e'/h)$. Consider further that, arguably, both $P(e_{m+1}/h \,\&\, e_1 \,\&\, e_2 \,\&\, e_3 \,\&\ldots\&\, e_m)$ and $P(e_{m+1}/-h \,\&\, e_1 \,\&\, e_2 \,\&\, e_3 \,\&\ldots\&\, e_m)$ tend to 1 as more and more evidence supportive of h is generated, which means that the ratio $P(e_{m+1}/h \,\&\, e_1 \,\&\, e_2 \,\&\, e_3 \,\&\ldots\&\, e_m)/P(e_{m+1}/-h \,\&\, e_1 \,\&\, e_2 \,\&\, e_3 \,\&\ldots\&\, e_m)$ tends to 1 as well (or at least greater than 1, depending on how one assesses the impact of $-h$). It follows that (1c) will always hold and that it is never of any epistemic value to switch from E to E'. In other words, the prescription to change observational procedures, as per the demand of robustness, fails to hold when the experiment to which one might switch is of sufficiently poor quality—a result that seems intuitively right.

This objection to robustness might be readily admitted by robustness advocates, who could then avert the problem by requiring that the observational procedures we are considering meet some minimal standard of reliability (the approaches of Bovens and Hartmann 2003 and Sober 2008, discussed below, include this requirement). So, for example, we might require that $P(e'/h) > P(e'/-h)$ (i.e., if h entails e', E' to some minimal degree tracks the truth of h), so that as the left side of (1c) tends to 1 we will be assured that there will a point where it is wise to switch to E'. But let us consider a situation where E' is such that $P(e'/h) = .0002$ and $P(e'/-h) = .0001$ (note that such an assignment of probabilities need not be inconsistent; it may be that for a vast majority of time, E' does not produce any report at all). In due course it will then become advisable on the positive relevance criterion to switch from E to E', even where $P(e/h)$ is close to 1 (i.e., where E is highly efficient at tracking the truth of h as compared to E', which is quite weak at tracking the truth of h). In fact, let $P(e/h) = .9$ and $P(e/-h) = .5$ (here, E would be particularly liberal in generating e). It follows that $P(e/h)/P(e/-h) = .9/.5 = 1.8$ and $P(e'/h)/P(e'/-h) = .0002/.0001 = 2$, and thus with just one trial h is better supported by a confirmatory result from experiment E' than from E. This seems very unintuitive. Given how poor E' is at tracking the truth of h—with one trial, generating e' is for all practical purposes as unlikely given h as with $-h$ (i.e.,.0001 $\approx$.0002)—E should stand as a better experiment for testing the truth of h, most certainly at least with one trial. Perhaps after 100 or so trials E' might be a valuable experiment to consider. But then we have the contrary consideration that, if the probabilistic independence between the outcomes of E and E' fails to hold, the right side of (1c),

$$\frac{P(e'_j/h \,\&\, e_1 \,\&\, e_2 \,\&\, e_3 \,\&\ldots\&\, e_m)}{P(e'_j/-h \,\&\, e_1 \,\&\, e_2 \,\&\, e_3 \,\&\ldots\&\, e_m)}$$

also approaches 1 with more trials, making E' less and less attractive as compared to E.

What we have found so far, then, is that incorporating considerations of experimental reliability into the Bayesian formalism complicates the assessment that it is beneficial to the confirmation of a theoretical hypothesis to switch observational procedures. However, the problem may not be so much Bayesianism as it is the way we have modified Bayesianism to accommodate the uncertain reliability of observational processes. Notably, consider how one may go about evaluating the left side of (1c),

$$\frac{P(e_{m+1} / h \& e_1 \& e_2 \& e_3 \& \ldots \& e_m)}{P(e_{m+1} / -h \& e_1 \& e_2 \& e_3 \& \ldots \& e_m)}$$

We have assumed that h entails e but that, given a less than perfectly reliable observational process, $1 > P(e_i/h) > 0$. How then does one evaluate the denominator, $P(e_{m+1}/-h \& e_1 \& e_2 \& e_3 \& \ldots \& e_m)$? We might suppose that $P(e/-h)$ is low relative to $P(e/h)$ (otherwise, experiment E would be of little value in confirming h). For simplicity, let $P(e/-h)$ be close to zero. As data confirmatory of h come streaming in, $e_1, e_2, e_3, \ldots e_m$ and so on, we have said that $P(e_{m+1}/-h \& e_1 \& e_2 \& e_3 \& \ldots \& e_m)$ will approach unity. But is that so given the conditional assumption $-h$? One might legitimately say that $P(e_{m+1}/-h \& e_1 \& e_2 \& e_3 \& \ldots \& e_m)$ remains unchanged, since the objective probability that an observational procedure generates a data report e given the assumption $-h$ does not vary with the state of the evidence (though of course one's subjective probability may vary). So, with $P(e/-h)$ starting out near zero, $P(e_{m+1}/-h \& e_1 \& e_2 \& e_3 \& \ldots \& e_m)$ remains near zero, and the left side of (1c) remains high, with the result that it would be perennially preferable to stay with E.

In fact, a similar problem of interpretation afflicts the numerator as well, though it is less noticeable since $P(e/h)$ starts out high to begin with (given that we have an experiment that is presumably reliable and presumably supportive of h). And, we might add, this problem attends Franklin and Howson's formalism described above. In their Bayesian calculation, they need to calculate $P(e_1 \& e_2 \& e_3 \& \ldots \& e'_{m+1}/h)$. Where $P(e/h) = 1$, and both E and E' are perfectly reliable experiments, $P(e_1 \& e_2 \& e_3 \& \ldots \& e'_{m+1}/h) = 1$ as well. However where $P(e/h) < 1$, the value of $P(e_1 \& e_2 \& e_3 \& \ldots \& e'_{m+1}/h)$ becomes less clear, for the reasons I have given: on the one hand (subjectively), we grow to expect evidence e_i and so $P(e_1 \& e_2 \& e_3 \& \ldots \& e'_{m+1}/h)$ increases; on the other hand (objectively), $P(e_1 \& e_2 \& e_3 \& \ldots \& e'_{m+1}/h)$ remains close to the initial value of $P(e_i/h)$.

Perhaps then our recommendation should be to attempt a different approach to incorporating into Bayesianism considerations of observational reliability. A decade after their first approach, Franklin and Howson suggested a different Bayesian formalism that respects the less than perfect reliability of observational processes. Specifically, Howson and Franklin (1994) propose to revise the formalism to accommodate the 'reliability' factor in the following way. They consider a case where

> we have a piece of experimental apparatus which delivers, on a monitor screen, say, a number which we interpret as the value of some physical magnitude m currently being measured by the apparatus. We have a hypothesis *H* which implies, modulo some auxiliary assumptions *A*, that *m* has the value *r*. Hence H implies that if the apparatus is working correctly *r* will be observed on the screen. Let us also assume that according to the experimenter's best knowledge, the chance of *r* appearing if *H* is true but the apparatus is working incorrectly is so small as to be negligible. On a given use of the apparatus *r* appears on the screen. Call this statement *E*. Let *K* be the statement that the apparatus worked correctly on this occasion. (461)

Under these conditions *H* and *K* entail *E*. We assume, moreover, that *H* and *K* are probabilistically independent. Then, by Bayes' theorem (keeping Howson and Franklin's symbolism),

$$P(H/E)=\frac{P(H)[P(E/H\,\&\,K)P(K/H)+P(E/H\,\&\,-K)P(-K/H)]}{P(E)}$$

Since, given our assumptions, $P(E/H\&K) = 1$, $P(E/H\&{-K}) = 0$ (approximately) and $P(K/H) = P(K/{-H}) = P(K)$ (probabilistic independence), it follows that

$$P(H/E)=\frac{P(H)P(K)}{P(E)} \qquad (2)$$

This equation, Howson and Franklin claim, 'summarizes the intuitively necessary result that the posterior probability of H on the observed experimental reading is reduced proportionally by a factor corresponding to the estimated reliability of that reading' (462; italics removed), where this estimated reliability is denoted by $P(K)$. This is an innovative approach, but it is unclear whether it generates the right results.

Suppose we have an observational process designed to produce data signifying some empirical phenomenon but that, in fact, is completely irrelevant to such a phenomenon. For example, suppose we use a thermometer to determine the time of day or a voltmeter to weigh something. The generated data from such a process, if used to test theoretical hypotheses, would be completely irrelevant for such a purpose. For example, if a hypothesis (H) predicts that an event should occur at a certain time (E), checking this time using a thermometer is a very unreliable strategy, guaranteed to produce the wrong result. As such, our conclusion from such a test should be that the hypothesis is neither confirmed nor disconfirmed—that is, $P(H/E) = P(H)$. But this is not the result we get using Howson and Franklin's new formalism. For them, an experiment is highly unreliable if the apparatus fails to work correctly and a thermometer completely fails to record the time. As such, $P(K) = 0$, from which it follows from (2) that $P(H/E) = 0$. In other words, on Howson and Franklin's account, the thermometer 'time' reading disconfirms the hypothesis (assuming $P(H) > 0$), whereas it should be completely irrelevant. What this means is that we cannot use the Howson and Franklin approach to adequately represent in probabilistic terms the reliability of observational procedures and so cannot use this approach in probabilistically assessing the value of robustness reasoning.

In 1995 Graham Oddie (personal correspondence) proposed a different approach to incorporating into the Bayesian formalism the matter of experimental reliability, taking a clue from Steve Leeds. He suggests we start with an experimental apparatus that generates 'readings,' R^E, indicating an underlying empirical phenomenon, E. Oddie assumes that our only access to E is through R and that the experimental apparatus produces, in addition to R^E, the outcome R^{-E} indicating $-E$. He then formalizes how confident we should be in H, given that the experiment produces R^E, as follows:

$$P(H/R^E) = P(H\&E/R^E) + P(H\&-E/R^E)$$

$$= P(H/E\&R^E)P(E/R^E) + P(H/-E\&R^E)P(-E/R^E)$$

He then makes the following critical assumption: We assume the apparatus we are using is a 'pure instrument' in the sense that its power to affect confidence in H through outputs R^E and R^{-E} is purely a matter of its impact on our confidence in E. In other words, E and $-E$ override R^E and R^{-E}. This is just to say that $P(H/E \,\&\, R^E) = P(H/E)$ and $P(H/-E \,\&\, R^E) = P(H/-E)$. This gives us the key equation,

$$(\mathrm{OL})\ P(H/R^E) = P(H/E)P(E/R^E) + P(H/-E)P(-E/R^E)$$

(OL stands for Oddie–Leeds), which Oddie argues is the best way to update our probability assignments given unreliable evidence. Note that with Oddie's formalism, we are able to generate the right result if the apparatus is maximally reliable—if $P(E/R^E) = 1$, $P(H/R^E) = P(H/E)$—and also if R^E is irrelevant to E—if $P(E/R^E) = P(E)$ and $P(-E/R^E) = P(-E)$, then $P(H/R^E) = P(H)$—the place where the Howson and Franklin's (1994) formalism fails.

What does (OL) say with regards to the value of robustness? Let us consider two observational procedures that generate, respectively, readings R and R', both of which are designed to indicate the empirical phenomenon E (we drop superscripts for simplicity). Thus we have the equations

$$P(H/R) = P(H/E)P(E/R) + P(H/-E)P(-E/R)$$

$$P(H/R') = P(H/E)P(E/R') + P(H/-E)P(-E/R')$$

from which we can derive

$$P(R/H) = P(E/H)P(R/E) + P(-E/H)P(R/-E) \tag{3a}$$

$$P(R'/H) = P(E/H)P(R'/E) + P(-E/H)P(R'/-E) \tag{3b}$$

respectively. It can then be independently shown that

$$P(H/R) > P(H/R') \text{ iff } P(R/H)/P(R/-H) > P(R'/H)/P(R'/-H) \tag{4}$$

From (3a), (3b) and (4), it follows that

$$P(H/R) > P(H/R') \text{ iff } P(R/E)/P(R/-E) > P(R'/E)/P(R'/-E) \quad (5a)$$

(see Appendix 3 for proof). This biconditional has a clear similarity to our first attempt to incorporate issues of reliability into Bayesian confirmation theory; recall (1b):

$$P(h/e) > P(h/e') \text{ iff } P(e/h)/P(e/-h) > P(e'/h)/P(e'/-h)$$

The difference is that the meaning of $P(R/E)$ is clearer than that of $P(e/h)$. Whereas the latter is a mixture of causal and theoretical factors in the way I am interpreting it, the former has arguably a simpler meaning: With an observational process that generates a reading R, how well does this process thereby track the empirical phenomenon E? But the benefit stops there once we consider multiple repetitions of this process. Suppose we generate a series of readings $R_1, R_2, \ldots, R_n$ from the first observational procedure. At what point is it beneficial to halt this collection of readings and begin collecting readings from the other procedure, which generates the series $R'_1, R'_2, \ldots, R'_n$? Let us turn to (5a); we derive a biconditional that is reminiscent of (1c): $P(H/R_1 \,\&\, R_2, \ldots, R_{m+1}) > P(H/R_1 \,\&\, R_2, \ldots, R'_j)$ (i.e., R_{m+1} better confirms H than R'_j, after having witnessed a series of results $R_1 \,\&\, R_2, \ldots, R_m$) if and only if

$$\frac{P(R_{m+1}/E\& R_1 \& R_2,\ldots,R_m)}{P(R_{m+1}/-E\& R_1 \& R_2,\ldots,R_m)} > \frac{P(R'_j/E\& R_1 \& R_2,\ldots,R_m)}{P(R'_j/-E\& R_1 \& R_2,\ldots,R_m)} \quad (5b)$$

Like (1c), (5b) suffers (analogous) problems. Notably there is the question of interpretation. Suppose that $P(R/E)$ is relatively high—the observational procedure is efficient at generating readings that indicate a phenomenon E, when E is present—and that $P(R/-E)$ is relatively low—the procedure seldom produces 'false positives'. Suppose further that this procedure generates a string of positive readings, $R_1, R_2, \ldots, R_m$. What value should we give to $P(R_{m+1}/-E \,\&\, R_1 \,\&R_2, \ldots, R_m)$?

On the one hand we expect it to be low, when we consider the condition $-E$; on the other hand, we expect it to be high, when we consider the track record of $R_1, R_2, \ldots, R_m$. So the Oddie–Leeds formalism, despite making clear in probabilistic terms the reliability of observational data, still suffers from a lack of clarity when it comes to assessing the impact of repeated trials on the confirmation of a hypothesis. Without that clarity, there's no point in using this formalism to either support or confute the value of robustness in establishing the reliability of an observational procedure.

In contrast to the Bayesian approaches to defending robustness that we have examined thus far, a straightforward, likelihoodist justification of robustness can be found in Sober (2008, 42–43). The case study Sober uses to illustrate his argument involves two witnesses to a crime who act as independent observers. We let proposition P stand for 'Sober committed the crime,' and $W_i(P)$ stand for 'witness W_i asserts that P'. Sober further imposes a minimal reliability requirement:

$$\text{(S)} \quad P[W_i(P)/P] > P[W_i(P)/-P], \text{ for } i = 1,2$$

He then asks: Where we have already received a positive report from one of the witnesses regarding P, is the confirmation of P enhanced by utilizing a positive report from the other witness? Given the likelihoodist perspective from which Sober (2008) works,

> observations O favor hypothesis H_1 over hypothesis H_2 if and only if $P(O/H_1) > P(O/H_2)$. And the degree to which O favors H_1 over H_2 is given by the likelihood ratio $P(O/H_1)/P(O/H_2)$. (32)

Obviously what we have in (1b), and in a modified form in (5a), is a comparison of such likelihood ratios from different observational procedures, and indeed Sober takes an approach in comparing observational procedures that is similar to what we have suggested. He asks us to consider the relevant likelihood ratio in a case in which we retrieve reports from independent witnesses and to compare that case to a different sort of case where we advert solely to the testimony of one witness. The details as he works them out are as follows: for independent witnesses W_1 and W_2,

$$\frac{P[(W_1(P)\&W_2(P))/P]}{P[(W_1(P)\&W_2(P))/-P]}=\frac{P[(W_1(P)/P)]}{P[W_1(P)/-P]}\cdot\frac{P[(W_2(P)/P)]}{P[W_2(P)/-P]} \quad (6a)$$

Since by (S) the ratios on the right being multiplied are each greater than one, it follows that the ratio on the left is larger than one and larger than each of the ratios on the right. From here he concludes that his likelihoodism is able to

> reflect the common sense fact that two independent and (at least minimally) reliable witnesses who agree that *P* is true provide stronger evidence in favor of *P* than either witness does alone. (42–43)

One might think that there is something wrong with (6a) in that, given the first witness has testified to the truth of *P*, the second ratio on the right side should be

$$(*) \quad \frac{P[(W_2(P)/P\&W_1(P)]}{P[(W_2(P)/-P\&W_1(P)]}$$

However, Sober claims that the right side of (6a) is correct given the independence of the witnesses, which he calls 'independence conditional on the proposition reported: $P[(W_1(P)\ \&\ W_2(P))/P] = P[(W_1(P)/P]$ $P[(W_2(P)/P]$' (2008, 42, footnote 22; italics removed). He doesn't believe there is an unconditional independence between the testimonies of reliable witnesses—we'd expect that $P(W_2(P)/W_1(P)) > P(W_2(P))$. In other words, learning *P* (or –*P*) screens off the impact learning $W_1(P)$ might have on our assessment of the probability of $W_2(P)$. But if this is true for $W_2(P)$, then it is true for $W_1(P)$ as well, for by Sober's 'independence conditional on the proposition reported' criterion, $W_1(P)$ is independent of $W_1(P)$ just as it is independent of $W_2(P)$: *P* (or –*P*) screens off the impact learning $W_1(P)$ might have on our assessment of the probability of $W_1(P)$ just as it does with

$W_2(P)$. It is irrelevant that $P(W_1(P)/W_1(P)) > P(W_1(P))$ since that is the separate matter of the unconditional dependence between the testimonies of reliable witnesses. By comparison, the value of $P[(W_i(P)/P]$ is unaffected by retrieving the same witness report twice over. Thus we should have

$$\frac{P[(W_1(P)\,\&\,W_1(P))/P]}{P[(W_1(P)\,\&\,W_1(P))/-P]} = \frac{P[(W_1(P)/P]}{P[(W_1(P)/-P]} \cdot \frac{P[(W_1(P)/P]}{P[(W_1(P)/-P]} \qquad (6b)$$

and so, by parity of reasoning, attending to the first witness's positive report a second (and a third, and a fourth...) time gives us a stronger confirmation again and again. Nor can we fix the problem by using (*) instead since, analogously to a problem we cited above for (1c) (and for 5(b)), it is difficult to know how to evaluate (most especially) the denominator, $P[(W_2(P)/-P\&W_1(P)]$: The $-P$ tends to diminish the value we give to the probability of $W_2(P)$, whereas the $W_1(P)$ tends to increase it. So Sober's argument for the value of retrieving independent reliable witness reports, as opposed to sticking with just one witness, breaks down at its most crucial point.

The last probabilistic approach we consider derives from Bovens and Hartmann (2003). Bovens and Hartmann provide a highly complex Bayesian justification for robustness, one that is strongly motivated by comments made by C. I. Lewis in 1946. Lewis claims that, where we receive multiple independent witness reports that converge, we should be inclined to regard these reports as approaching truthfulness since, '[o]n any other hypothesis than that of truth-telling, this [convergence] is highly unlikely' (346; quoted in Bovens and Hartmann 2003, 56). Clearly, Lewis is advancing a version of the no-miracles argument that we critiqued above. In Bovens and Hartmann's (2003) hands, however, this argument becomes more subtle: Instead of assuming dichotomously that an observational process is reliable or not (an assumption that earlier led to a dilemma), they assume that a process is reliable to varying degrees. More precisely, they

> assume that if witnesses are not reliable, then they are like randomizers. It is as if they do not even look at the state of the world to

> determine whether the hypothesis is true, but rather flip a coin or cast a die to determine whether they will provide a report to the effect that the hypothesis is true. (57)

In their formalism, Bovens and Hartmann let REL stand for the assertion that an observational process (a 'witness') is reliable and incorporate into their proofs the probability value *P*(REL). For a witness who is completely unreliable, *P*(REL) = 0, which means that the witness is a randomizer who sometimes asserts observation reports that are right regarding the truth of a hypothesis and sometimes asserts reports that are wrong, all in a random manner. On the other hand, where *P*(REL) = 1, the witness's reports are consistently correct. In between these values, the witness is reliable to some intermediate degree in the sense of having some tendency to assert true reports, even if only slightly if *P*(REL) is just above zero. In other words, the situation where a witness systematically gets the wrong answer (is 'antireliable') is not factored into Bovens and Hartmann's account. This omission is quite significant for their argument, since people could be unreliable not in the sense that they are only randomly right but instead are systematically wrong all of the time. Thus, because of the way Bovens and Hartmann have set up their formalism, where a value of *P*(REL) above zero means that a witness has at least some small positive tendency to issue correct reports, the task of defending robustness is significantly lightened.

Given their omission of the antireliable case, Bovens and Hartmann are able to construct a fairly convincing case in support of Lewis's robustness intuition. To my knowledge, it is one of the more compelling probabilistic arguments for robustness that one can find, though it does suffer a critical flaw (as I will argue). Fortunately we can express the Lewisonian intuition underlying the Bovens and Hartmann approach without delving into their calculational details (the interested reader is invited to consult Bovens and Hartmann 2003, 60–66). Suppose we have a group of independent witnesses reporting on some empirical phenomenon where each witness is minimally reliable (i.e., to perhaps only a small degree, each witness has a tendency to truthfully report on the phenomenon). Suppose, moreover, that the witnesses are unanimous in their reports about this phenomenon. There is then a convincing case to say that the probability of this report being true increases, given this convergence among witnesses (assuming there are no dissenters), more so than if we had recorded the testimony of the same witness's testimony repeatedly. This probabilistic increase exhibits the extra confirmatory boost that is afforded by robustness reasoning.

Of course, this argument only goes through if the witnesses are independent. For example, the argument fails if there is collusion among the witnesses or an extraneous common cause for their convergence of opinion. The matter of the independence of witnesses, or of different empirical reports, generally speaking, is a subject of some controversy and is probably not formalizable in logical terms. To get a sense of the difficulty, consider the analysis of 'independence' introduced by Bovens and Hartmann (2003) that is fundamental to their proof of the value of robustness:

> The chance that we will get a positive report from a witness is fully determined by whether that witness is reliable and by whether the hypothesis they report on is true. Learning about other witness reports or about the reliability of other witnesses does not affect this chance. (61)

The reader will readily note the similarity of this approach to Sober's notion of 'independence conditional on the proposition reported': Just as with Sober's approach, once we assume the truth of the claim being reported on, the chance that a witness report is true is unaffected by the presence of other (positive or negative) witness reports. The main difference between Sober's approach and the Bovens–Hartmann (B–H) approach is that the latter conditionalizes as well on how reliable the witness is, whereas the former includes only a minimal reliability requirement. Nevertheless, the B–H approach stumbles at the same place as Sober's approach: Where for Sober a witness report $W_1(P)$ is independent of $W_1(P)$ just as it is independent of $W_2(P)$, so for Bovens and Hartmann 'a positive report from a witness,' symbolized by REP_1, is independent of REP_1 just as it is independent of REP_2 (a positive report from a second witness) since 'the chance that we will get a positive report from a witness is fully determined by whether that witness is reliable and by whether the hypothesis they report on is true' (61), not on whether that report has already been given. As such, we have with the B–H approach the same regrettable result we have for Sober's approach: that retrieving a witness's report again and again would succeed in enhancing the confirmatory power of this report.

One way we might diagnose what is going wrong with the Sober and B–H approaches is to point out that they are attempting to work with an objective notion of probability as a way of maintaining the independence of witness (or empirical) reports. This becomes especially clear with Bovens and Hartmann in their assessment of the conditional probability of a witness report REP given the truth of the hypothesis under test (HYP) along with the assumption that the witness is reliable REL, which they assess as

$$P(\text{REP}/\text{HYP}, \text{REL}) = 1$$

This equation would be accurate if we understood the probability of the witness report to be the objective chance for this report to be true. But with robustness, one might argue, what we should be considering instead is the subjective probability that we attach to a witness report, which may vary from the objective probability of the report, especially if one lacks an awareness of both the truth of the hypothesis being reported on and the reliability of the witness. The subjective probability may well be more appropriate here since it gives the right result when a witness simply repeats a report: A report once given is assigned a probability of one, and so no further confirmation via conditionalization would be forthcoming. Moreover, once a report is given, it subjectively seems more likely that one would find analogous reports produced by other witnesses. To adapt Sober's terminology, there is an unconditional *dependence* between the testimonies of (presumably) reliable witnesses. But in that case, of course, we lose the independence that is supposed to be the trademark of robustness reasoning, if this independence is to be understood in probabilistic terms: The (subjective) probability of a witness report that is repeated by different witnesses increases just as it does with repeated reports by a single witness. And, in fact, isn't this what we should expect if we take the reports of the other witnesses to bear on the truth of the claim under consideration? If other people are conveying positive witness reports about some claim and we take them to be at least minimally reliable, then we assess the likelihood that we would, similarly situated, convey the same positive report as to some degree increased. So the lesson we might derive here is that, to understand the independence that underwrites robustness reasoning, we need to comprehend this independence in a nonprobabilistic way.

We have, of course, other ways of understanding the independence of observational procedures—by simply thinking of them as utilizing different physical processes or, alternatively, as involving different theoretical assumptions (i.e., 'epistemic independence'). With these less precise but still suggestive interpretations of independence, the defense of robustness perhaps has force in just the nontechnical way Lewis suggests. We can put the idea—an elaboration of the core argument for robustness—in the following way. If an observational report is generated by two (or more) distinctly different physical process (or as a product of two or more distinct theoretical assumptions), then we reduce the chance that the report is simply an artifact of one of these process (or one of these assumptions) since it is unlikely that the *same* artifact could be independently produced. In other words, it is not the case that one or other of the processes (or one or other of the assumptions) is uniquely responsible for ensuring the production of this report; the production of the report is not the result of a physical (or theoretical) bias informing some particular observational procedure. Consequently, there must be some other explanation for this production, presumably the reliability of all the processes that generate this report, along with the presumption that the report is true. This is the sort of insight that, I believe, drives the proponents of robustness—and we concede its intuitiveness.

Of course, the question remains whether this insight is valid, a matter I defer until chapter 6. Whereas we have so far been construing independence as involving distinct physical processes, there is the matter of interpreting independence as involving distinct theoretical assumptions (epistemic independence). I deal with this approach to robustness at the end of this chapter. In advance of that discussion, let's examine a different, though highly influential, approach to defending robustness reasoning, a pragmatic approach initially formulated by William Wimsatt and subsequently elaborated by Kent Staley.

PRAGMATIC APPROACHES TO ROBUSTNESS

In an oft-cited paper, Wimsatt (1981) provides the following argument on behalf of the value of robustness reasoning. To begin with, he utilizes a distinction drawn by Richard Feynman between 'Euclidean' and 'Babylonian' ways of structuring a theory (128–130). Euclidean theoretical structures are such that there is a relatively small core of axioms from which all the remaining statements of a theory can be derived. Thus, for each theoretical statement, there is one definitive, unique line of reasoning that justifies it. Babylonian structures, by contrast, are more diverse in how theoretical claims are justified; there are a variety of ways, each involving different assumptions, by means of which theoretical claims are justified. Feynman, as Wimsatt (1981) recounts his views, defends the use of Babylonian structures for theories on the grounds that physical laws as a consequence are multiply derivable and so enjoy more stability despite the occurrence of theory change. By being independently derivable, a bulk of a theory may change and yet a physical law will remain since it is derived from other parts of the theory that have persisted. Such multiple derivability, on Wimsatt's view, 'not only makes the overall structure [of a theory] more reliable' but also allows us to identify those theoretical laws that are 'most robust and . . . [so] most fundamental' (130). By comparison, the rationale for Euclidean structures, on Wimsatt's view, is

> to make the structure of scientific theory as reliable as possible by starting with, as axioms, the minimal number of assumptions which are as certain as possible and operating on them with rules which are as certain as possible. (131)

So both strategies, the Babylonian and the Euclidean, have as their intent to secure the reliability of a theory. The question is: Which succeeds better?

For Wimsatt, our preference should be for Babylonian (i.e., robust) structures for the following reason. For a theoretical claim to be justified in a Euclidean structure, there is a singular line of reasoning stemming from the fundamental axioms to the claim in question. Now, each assumption and each inferential step in this line of reasoning will have some probability of being in error (either the assumption will have a certain chance of being false or the inferential step will have a certain probability of failing), and the string of all these assumptions and steps of reasoning, put in an order that captures the derivation of a theoretical claim, will compound these probabilities of error. As a result, a serial proof of a theoretical hypothesis from a limited, beginning set of axioms has a higher chance of failure (given the probabilistic independence of each component step/assumption) than that of any particular assumption or inferential step. Conversely, when one derives a theoretical claim in a variety of ways, as one would with a Babylonian theory, each of these ways will have some chance of success (i.e., 1 – the chance of failure); and if each of these alternative derivations is independent of one another, the overall chance of success is the sum of all these chances of success, where this sum will be larger than the chance of success for the most likely, successful derivation. So, as Wimsatt (1981) summarizes his argument, 'adding alternatives (or redundancy, as it is often called) always increases reliability, as von Neumann . . . argued in his classic paper on building reliable automata with unreliable components' (132–133; see 131–134 for the fuller presentation of this argument).

This is a fascinating probabilistic argument that has the further benefit of explaining why theories with inconsistencies are still usable: An inconsistency need not afflict the whole theory (as it would with a Euclidean structure) but only certain independent lines of reasoning. Yet, despite these benefits, Wimsatt's reasoning is in fact irrelevant to the issue of robustness with regard to experimental procedures. We can frankly admit that if we are designing a machine to perform a task, then it is helpful to have backup systems in place that will undertake this task if a primary system fails. But reliability in this sense is not an epistemic notion but a pragmatic one. By 'pragmatic reliability,' what is sought is not specifically a system that generates truthful results. What is sought is a system that generates consistent results, results that have a high probability of being generated again and again, whether or not it is a result that expresses a truth. With this meaning of 'reliability,' we can say that a car is 'reliable' in that it is guaranteed to start and run. We could even say that a machine is reliable if it is designed to produce, in a consistent manner, false claims.

But clearly this is not a notion of reliability that is relevant to the epistemic appraisal of experimental set-ups.

To illustrate the sort of problem I have in mind, suppose we have three independent experimental tests for the existence of a certain phenomenon, and suppose each test has a 50% chance of recording the existence of this phenomenon, whether or not the phenomenon is present. That is, each test is an unreliable indicator of this phenomenon; its results are completely randomly connected to the state of the world. Still, it is nevertheless the case that, taken together, the overall chance of at least one of the tests recording a positive indicator of the phenomenon is almost 90% (eight possible combinations of results for the three tests, seven of which involve at least one test yielding a positive result). So we have a fairly high success rate in generating an indicator of the phenomenon, due to the robustness of our methodology. It is as if we were trying to generate a positive report regarding the phenomenon and so build into our experimental regime 'redundant' indicators for this phenomenon, in case some of the tests don't produce a positive result. But surely we do not generate thereby a result that has epistemic significance. There is no guarantee that this redundancy will emanate in a truthful report—only a guarantee that a certain kind of report will (most) always be generated.

A similar objection can be raised regarding Feynman's preference for Babylonian theoretical structures. Robust physical laws in such structures have multiple derivations that can assure the justified persistence of such laws despite theory change. But what if each of these derivations is riddled with inaccuracies, flawed assumptions and invalid inferences? In such a case, the multiple derivability of a law would be irrelevant to the epistemic merit of this law. Ultimately, we need derivations that meet certain standards of reliability (such as relying on true assumptions as well as involving inferences that are either inductively or deductively cogent), not simply derivations that converge in their assessments, leaving aside the question of the epistemic legitimacy of these derivations.

It is true that the form of robustness to which Wimsatt (1981) is referring in his probabilistic argument is more akin to inferential robustness than to measurement robustness (as these terms are defined in the Introduction). The robustness he cites attaches to claims (i.e., laws) that are multiply derivable, not multiply generated using different observational procedures. But Wimsatt is in fact somewhat imprecise in his presentation of robustness notions. To him, 'robustness analysis' forms a 'family of criteria and procedures' (126) that is instantiated in a variety of contexts, and apart from the above argument from Euclidean and Babylonian theoretical structures, he provides no other sustained arguments for why robustness is to be valued. For instance, in specifically discussing the robustness of observational procedures, he highlights a case where 'the boundaries of an ordinary object, such as a table, as detected in different sensory modalities (visually, tactually, aurally, orally), roughly coincide, making them robust,' and in answering the question why this robustness 'is ultimately the primary reason why we regard perception of the object as veridical rather than illusory,' he provides the one-sentence explanation, 'it is a rare illusion indeed which could systematically affect all of our senses in this consistent manner' (144). This again is the core argument for robustness, put in very succinct form, which I address in chapter 6. So it is not unreasonable to think that, for Wimsatt, his more extensive, pragmatic argument for robustness applies to measurement robustness as well and to the other seven instantiations of robustness reasoning that he identifies (126–127).

But even if it is not clear that Wimsatt intends his pragmatic argument to apply to cases of measurement robustness, a version of his argument is indeed so applied by Kent Staley (2004). On Staley's approach there are two possible kinds of benefits with robust evidence. We assume, to begin with, a procedure that leads to an observational claim. The first benefit he cites emanates from the fact that one might identify sources of empirical support for the procedure itself, which can serve to put this observational claim on firmer footing. As Staley describes this option, where the results of the original observational procedure 'are considered as first-order evidence for the primary hypothesis,' the results of a different observational procedure provide 'evidential support for assumptions about the first [procedure] on which that evidence claim rests' (474). Strictly speaking, though, this is not a case of robustness reasoning: This is performing a test on an observational procedure to ensure its good functioning (an example of what we later call 'targeted testing'). There is no sense here in which independent observational procedures are found to converge on the same observed result, since the separate procedures—the original and the procedure that serves to test it—issue in different observed results. By comparison, the second benefit Staley cites with robust evidence is to the point. He describes this benefit as follows:

> [the] use [of] convergent results from a second test... serve as a kind of 'back up' evidence against the possibility that some assumption underlying the first test should prove false. The difference is similar to the following. An engineer has a certain amount of material with which to construct the pilings for a bridge. Calculations show that only 60% of the material is needed to build a set of pilings sufficient to meet the design specifications, but the extra material, if not used, will simply go to waste. The engineer decides to 'overengineer' the pilings with the extra material [and]... use the extra material to produce additional pilings.... Like the engineer who chooses to build extra pilings, the scientist might use convergent results to [serve as]... a kind of back-up source of evidence that rests on different assumptions than those behind the primary evidence claim. [As such] one might be protected against the failure due to a wrong assumption of one's claim about how strong the evidence is for a hypothesis. In effect, this is to claim that, although one's assumptions might be wrong, one's claim that the hypothesis has evidence of some specified strength in support of it would still be correct (though not for the reasons initially given). (474–475)

The benefit Staley here cites with robustness is clearly the same benefit to which Wimsatt refers, that of having multiple, redundant evidential supports for a (theoretical or observational) claim. And once more, just as we found with Wimsatt's approach, this benefit is purely pragmatic: We are ensuring that a claim has 'support' under diverse circumstances, without necessarily considering whether that support is epistemically meritorious.

Unfortunately, Staley seems unaware of the purely pragmatic nature of the benefit he ascribes to robustness (the first benefit he cites for robustness—evidential support for the assumptions that underlie an observational procedure—is clearly epistemic, but here he is not really talking about robustness). That lack of awareness aside, Staley sees himself as furthering the discussion on robustness by identifying and responding to various criticisms that can be launched against robustness. His claim is that these criticisms can be effectively rebutted if robustness is supplemented in the right sort of way. It is worthwhile examining these criticisms and Staley's responses as it permits us to deepen our understanding of when observational procedures can be said to be independent.

Staley (2004) starts by rightly noting that there are circumstances in which the epistemic significance of robustness is questionable. He cites two kinds of circumstances (472–473). First, there are cases of spurious convergence. In this sort of case, two independent empirical procedures generate similar results, but this is purely a matter of chance—one or other, or both, procedures have no reliable connection to the phenomenon under study, but through lucky happenstance they arrive at the same result. To illustrate, Staley offers the following example:

> Consider two particle detectors arranged as coincidence indicators, so that a particle passing through one will almost certainly pass through the other, producing two nearly simultaneous signals. Assume that two detectors are based on entirely different technologies and rely on different physical principles, so as to constitute independent means of detection, and that both detectors produce a signal at about the same time. The results satisfy the robustness requirement, being both convergent and produced independently. If, however, the second detector were so noisy that it had a 50% chance of producing a signal in the absence of any particle, we could safely conclude that the convergence of these independently produced results is without evidential value. (472)

The second sort of problem Staley cites for robustness involves a case where we have a concealed failure of independence. To use a nontechnical example, suppose we have two seemingly independent news sources (such as two different newspapers) producing the same story but only because each news source is being fed by the same correspondent. To argue on the basis of this convergence on behalf of the truthfulness of this story would be inappropriate. Similarly, we might have a situation where one empirical test is used to 'calibrate' another test (i.e., the results of one test guide the correctness of a second test) but this is unknown or forgotten. In this circumstance, two different procedures would generate the same result in a certain range of cases, but that would only be a product of the calibration, not a case of independently arriving at the same result. So, generally speaking, in both cases there is a failure in independence that leads to a convergence of results, and the observer who is unaware of this failure is mistakenly led to think that there is an extra presumption on behalf of the truth of these results because of their robustness.

These are important problems for robustness—the latter problem we are already familiar with as it relates to the definition and identification of independent observational procedures. What is of interest to us is how Staley suggests we can handle these problems: He suggests that we need to supplement robustness with a further methodological principle, that of 'discriminant validation,' according to which (on Staley's definition) we require 'different sources of evidence... [to] not yield convergent results when the phenomenon to be detected or measured is absent' (473). Discriminant validation is thus the converse of robustness, which sees different sources of evidence yielding convergent results when the phenomenon to be detected is present. By applying discriminant validation, Staley believes we can we arrive at the correct result in the problematic cases cited above. For example, as regards spurious convergence, Staley asserts:

> The results meet the requirements of convergent validation, but fail the test of discriminant validation. The second detector would frequently deliver such a confirming signal even if we employed it as an anti-coincidence detector. (474)

That is, in a case where the first detector (reliably) indicates the presence of a certain kind of incoming particle, and where we employ the second detector as an anticoincidence detector (and so use it to generate a positive result if some other kind of particle impacts the detector), the second detector might well fire, given how noisy it is, disconfirming the presence of the sought-for particle and thus refuting the initially claimed convergence of results. We thus have the benefit of discriminant validation: Whereas robustness can fool us where we have a spurious convergence, discriminant validation can serve to reveal where we have been mistaken.

Yet Staley's assessment of this case is uncertain. If a 50% chance of a convergent result is enough for us to worry that robustness (i.e., convergent validation) is giving the wrong result, then, since with an anticoincidence detector case we also have a 50% chance of making a correct discrimination (that is, the noisy detector fails to fire with a 50% chance, indicating the sought-for particle), we should worry that discriminant validation, too, is prone to give us the wrong result, in light of the spurious, imperfect nature of the detectors. But isn't the best way, in any event, to resolve this problem of spurious convergence to simply let the detectors run for a long enough time? In due course, given the chance nature of the second detector, repetitive trials will eventually reveal a lack of convergence defeating any (mistaken) robustness argument. Thus it is not clear that discriminant validation really adds much in cases of spurious convergence.

I am similarly skeptical about the value of discriminant validation in handling cases of failures of independence. It is true that discriminant validation can rule out these types of failures of independence. As regards the common cause (newspaper) case, where the relevant news story is false and the same correspondent feeds this information to both newspapers, the reports of these newspapers will again converge (here, expressing a false report). And, in the calibration case, where the calibrating procedure generates a false result, once more we will have a convergence of results involving both the calibrating procedure and the calibrated procedure. So applying discriminant validation to these cases apparently shows that they should be discarded despite their robustness, since 'different sources of evidence... yield convergent results when the phenomenon to be detected or measured is absent' (Staley 2004, 473). But that is only because the sources of information themselves (the news correspondent and the calibrating procedure) are inherently unreliable. Comparatively, what is objectionable about a case where two observational procedures always produce the same observed result when this result is accurate? In the newspaper case, where the correspondent is reliable, the issued stories of the two newspapers will converge whether or not the phenomenon under scrutiny is present or absent—the correspondent will reliably inform whichever is the case. And in the calibration case, if the calibrating procedure is reliable, both it and the calibrated procedure will generate truthful results, whether they report the occurrence of a phenomenon or its absence. Thus it is unclear what benefit is being provided by introducing the discriminant validation requirement. Why shouldn't 'different sources of evidence... yield convergent results when the phenomenon to be detected or measured is absent' (473)?

One might suggest here that Staley's definition of discriminant validation is flawed and needs revision. For example, one might revise it thus (as Staley does [personal correspondence]): Discriminant validation requires only that the sources of evidence not yield convergent *positive* results when the phenomenon is absent. This sounds like a fine principle: Surely we don't want experimental results to issue in positive results when the phenomenon to be measured is absent. This is as much to say that we want our testing schemes to be severe, in Deborah Mayo's sense (Mayo 1996). However, this new principle no longer looks much like the discriminant validation principle, as originally set forth by Donald Campbell and Donald Fiske (Campbell and Fiske 1959). As they define discriminant validation, one rules out tests if they exhibit 'too high correlations with other tests from which they were intended to differ' (81). Indeed, Staley (2004) provides just such a definition of discriminant validation—as he puts it, 'discriminant validation is a process of checking to see whether a particular process produces results that correlate too highly with the results of processes that should yield uncorrelated results' (474)—but does not make clear how this definition differs from his own, cited above. The Campbell and Fiske definition, we should emphasize, asks that tests not yield the same result when they should be generating different results, leaving aside the issue of whether the phenomenon to be measured is present or absent, and leaving aside whether the results are positive or negative. One of the classic cases where empirical inquiry fails discriminant validation, as recounted by Campbell and Fiske (1959), is the 'halo effect,' where, to take one example, one's initial perception of a person as having certain commendable traits influences one's attribution of further commendable traits to this person (see Campbell and Fiske 1959, 84–85—the term 'halo effect' was coined in Thorndike 1920). The underlying idea here is that our further attributions of traits to people should sometimes be expected to diverge somewhat from the traits we originally attributed to them and that we should be wary of cases where our attribution of traits is excessively consistent. Analogously, with regard to experimentation, subsequent experimental results should be expected to diverge somewhat from initial results; we should be wary of cases where experimental results are overly consistent with one another. Convergent and discriminant validation, seen in this way, thus contrast rather nicely: The first (robustness, or convergent validation) asserts that experimental results are more reliable if they agree with one another as retrieved using different physical procedures, whereas the second (discriminant validation) warns us about seeing too much consistency in our results when using different physical procedures.

So where does this leave Staley's modified version of the discriminant validation principle, which requires that sources of evidence not yield convergent, positive results when the relevant phenomenon to be detected or measured is absent? My assertion is that such a principle is unusable, for we would have to know beforehand whether the relevant phenomenon to be detected or measured really is absent, which would require having advance reliable knowledge about the phenomenon being investigated. Surely such knowledge is precisely the knowledge that is being sought in the first place by performing the experiments. By means of comparison, suppose the convergent validation (i.e., robustness) principle were to state that sources of evidence for a theory are more compelling when multiple independent tests yield the same convergent, positive result when the phenomenon to be detected or measured is present. This sounds like an excellent principle but for the fact that it is, also, unusable: To apply it we would need to know whether the phenomenon to be detected or measured is present to begin with, which is precisely the issue being investigated by the tests. Of course, one might reject the sort of argument I am providing here on the grounds that it treads too closely to the infamous experimenter's regress, made famous by the work of Harry Collins. Indeed, it is the sort of problem raised by the experimenter's regress. That regress concerns the attempt to prove that an experimental process is reliable by showing that it correctly identifies the phenomenon being investigated; the problem is that to determine that one has correctly identified the phenomenon being investigated, one needs to deploy the experimental process whose reliability is under scrutiny. Judgments of reliability thus seem to be locked in a justificatory circle. As it turns out, it is a clear merit of the robustness principle as we originally defined it that it avoids this sort of circularity objection. The point of robustness is to suggest that, when independent tests arrive at the same experimental result, this result is better justified than it would be if there were no such convergence—and however one appraises this reasoning (of course, I am doubtful about it), at least it does not require that we have identified the phenomenon being sought beforehand. Nor, indeed, does the discriminant validation principle formulated by Campbell and Fiske require such a prior awareness: One can ascertain that there are forms of expectation effects occurring in experimental practice without having settled the issue whether the practices are arriving at the right results about a hitherto unknown phenomenon. Thus, whereas the experimenter's regress is a potential problem for the discriminant validation principle proposed by Staley, which requires that sources of evidence not yield convergent, positive results when the relevant phenomenon to be detected or measured is absent, it is effectively averted by the robustness principle (as usually formulated) as well as by the discriminant validation principle as originally formulated by Campbell and Fiske (1959).

To summarize our assessment of Staley's critique of robustness, the spurious convergence problem Staley presents to us is not really a problem for robustness at all, since it can be handled in the usual case by simply collecting more evidence. On the other hand, the (concealed) independence problem is indeed a problem for robustness; sometimes it is difficult to know when observational procedures are independent and so difficult to tell whether robustness has an application. Staley attempts to use the principle of discriminant validation to manage this independence problem, but it turns out that the versions of discriminant validation he uses either lead us to the wrong result or are unusable because of problems analogous to those illustrated by Harry Collins's experimenter's regress. On the other hand, the original Campbell and Fiske (1959) version of discriminant validation has the ability to lead us to the right result, in that it can identify cases where differing observational procedures are not independent but biased; these are cases where the results of such procedures exhibit excessive consistency. Moreover, the Campbell and Fiske approach evades the experimenter's regress since its application does not require that one be aware beforehand of the true nature of the world. As a result, employing the notion of independence in the way elaborated by Campbell and Fiske puts us in a good position to exploit the merit contained in robustness reasoning. That merit, if it exists, is captured by what I called the 'core argument' for robustness (whose full appraisal occurs at the beginning of chapter 6). Comparatively, it is not captured by the pragmatic approaches suggested by Wimsatt and Staley, which fail to locate an epistemic merit for robustness. Nor do I think the value of robustness reasoning to be effectively captured by an alternative notion of independence, one set forth by Peter Kosso and others. We examine and critique this 'epistemic' notion of independence in the next section.

EPISTEMIC INDEPENDENCE APPROACHES TO ROBUSTNESS

As a way of elaborating on what we mean by the term 'independence,' an alternative conception of robustness suggests that we interpret the independence of observational procedures not as the independence of physical procedures but rather as the epistemic independence of the theoretical assumptions that underlie these procedures. One of the main proponents of this approach is Peter Kosso, who asserts that such independence is an aspect of Jean Perrin's famous arguments for the atomic hypothesis (arguments we examine in chapter 4). Kosso (1989) asserts,

> Perrin measured the same physical quantity in a variety of different ways, thereby invoking a variety of different auxiliary theories. And the reason that Perrin's results are so believable, and that they provide good reason to believe in the actual existence of molecules, is that he used a variety of independent theories and techniques and got them to agree on the answer. The chances of these independent theories all independently manufacturing the same fictitious result is small enough to be rationally discounted. (247)

In considering what is meant by 'independent theories' in this context, Kosso initially describes the idea as a sort of logical independence: Theories

T_1 and T_2 are independent of one another 'if our acceptance of T_1 as true (or rejection of T_1 as false) does not force us to accept T_2 as true (nor to reject T_2 as false)' (247). Subsequently, however, he expresses a preference for a different notion of independence 'which is more directly applicable to actual cases of objective testing,' a notion he calls 'independence of an account' (249). Here the idea is that, in testing a theory using observational results, we should avoid results that, in themselves, presuppose the truth of the theory under test.

Independence of an account is a highly popular approach to ensuring observational objectivity (proponents, in addition to Kosso, include Carrier [1989], Greenwood [1990]), Wylie [1990], and Sober [1999]). To understand why such a requirement might be necessary in ensuring the objectivity of observation, consider the following scenario. Suppose a creationist and an evolutionist are looking at a rock formation with the goal of identifying evidence for God's design. Both spot a fossil of what looks likes a reptile, one unlike any current reptile, and the creationist buoyed by her religious conviction announces that she clearly sees evidence of God's design by the intricate design God has inscribed in the rock. The evolutionist for his part argues that he has found, rather, evidence for evolution by the observation of an extinct, reptilian-looking ancestor to current life forms. Each observer, that is, examines the rock formation as filtered through his or her assumed theories and arrives at an observation that agrees with his or her theoretical preferences. In this way we can see that there are significant obstacles to objective theory testing if individuals are able to interpret the import of observations in accordance with their theoretical prejudices. Each of them, the creationist and the evolutionist, will be prone to a confirmation bias if each can filter observations through his or her preferred theoretical perspective. Thus the challenge of independence of an account—a challenge that can restore objectivity—is to ask each, the creationist and the evolutionist, to produce evidence for their respective views that doesn't presuppose their favoured theory. For example, the evolutionist might identify key morphological evidence that links the fossilized reptile with modern reptiles, whereas the creationist might introduce ecological evidence that no such 'real' reptiles could have existed at the current spot in the proposed geological time period.

An additional key feature of independence of an account is that it ties in well with robustness: If there is a concern about certain theoretical assumptions informing one's observations in a biased way, one can locate alternate ways of arriving at these observations that depend on different theoretical assumptions. By this means, we would then have shown that the observations do not presuppose the truth of the original set of assumptions. In other words, robustness can go toward ensuring independence of an account.

However, despite such considerations, I am uncertain whether independence of an account is needed for the purposes of ensuring objective observation. As I will show, in some cases it generates a false source of objectivity, and it can lead to unwise methodological advice that unduly restricts the work of scientific observers. Given all this, it follows that independence of an account cannot form the basis of an argument on behalf of the epistemic value of robustness.

To begin our assessment of independence of an account, let us consider more carefully what sort of theory of observation a proponent of independence of an account might have in mind. We begin by distinguishing two senses of observation (following Dretske 1969), the epistemic and the nonepistemic. According to the latter, observation is a nonconceptual relationship between an observer and a state of affairs, a relationship that is usually thought of as causal (although this is not necessary). With this view of observation, someone can observe a state of affairs so long as this state of affairs looks some way to the observer, even if the observer lacks the conceptual resources to recognize this state of affairs as looking this way. Many animals, we presume, observe things nonepistemically in light of their low levels of cognitive development, and I think we can affirm that all observers, whatever their intellectual acumen, observe things to some degree nonepistemically. For in advance of conceptualizing an observed object, one must be able to observe it in some manner, and only nonepistemic observation could play this role. Thus nonepistemic observation is an elemental feature of our experience of the world.

For most philosophers, though, observation is more importantly understood in its epistemic form. Epistemically speaking, to observe something is not just to have this thing appear to someone in some way—or in causal terms to be causally connected to a physical state of affairs through one's sensory organs. More than this, one must be in a position to conceptualize the object of observation. In this sense of observation, to observe is to 'observe that'. For instance, the creationist and evolutionist described above observed in a nonepistemic way the same rock formation, but they differed in regard to what they observed epistemically: Whereas the creationist observed *that* the rock contained God's designs, the evolutionist observed *that* there was a reptilian-looking ancestor to modern life forms fossilized in the rock. We might suggest, then, that the problematic circularity we have been citing in episodes of observation is a residue of an epistemic account of observation. With a nonepistemic account of observation, one is observing whatever it is that is causing one's observations, and this will be a fact unalterable by one's theoretical preconceptions. On the other hand, with epistemic observation, since what one observes is a by-product of what theories or concepts one brings to observation, we arrive at the problematic circumstance of observing what one thinks one is observing. For this reason one might suggest that, when we are considering epistemic observation, independence of an account may be a wise restriction, for by ruling out one's theoretical predictions as a filter on one's observations, one rules out the possibility of epistemically observing what one, theoretically, expects to observe.

After all, what other alternatives does one have here? One option might be to resort to observing the world purely nonepistemically. In such a case, there would not be any worry about preconceptualizing in a biased way the material of observation, since there is no conceptualization to start with. Yet, despite the resiliency of nonepistemic observation to the errors resulting from conceptual anticipation, it only succeeds at this task by draining observations of any propositional content. That is, nonepistemic observation strictly speaking does not 'say' anything to us and thus is quite useless at the task of theory testing, since what is tested are theoretical hypotheses that have, of necessity, a semantic dimension. Thus to have effective theory testing using nonepistemic observations, we need to reconfigure these observations in some way to make them 'epistemic,' which again leaves us susceptible to the hazard of interpreting one's observation in accord with one's favored theoretical preconceptions.

Another alternative is suggested by Jerry Fodor (1984). Fodor describes perceptual processes as composed of psychological modules whose functioning is partly inferential but also encapsulated in that the inferential elements of modules are mostly inalterable by higher cognition. Using an expression borrowed from Zenon Pylyshyn, Fodor describes the outputs to perceptual modules as 'cognitively impenetrable'. As we might put it, observations are epistemic, but their epistemic content is fixed to a significant degree by the modular mechanisms underlying perceptual processes. Fodor (1983) suggests that the original reason behind Pylyshyn's introduction of the phrase 'cognitively impenetrable' was to express the fact that an organism through observation 'sees what's there and not what it wants or expects to be there' (68). Less ambitiously, Fodor (1984) regards the cognitively implastic character of perception as essential to the objectivity of scientific observations in that it ensures the ability of theoretical opponents to reach a consensus about observations (42). Either way, the modularity of perception seemingly provides a way to counter the threat of relativity inherent in an epistemic view of observation and permits us to bypass recourse to the independence of an account approach.

Still, one might resist Fodor's modularity approach for the following reasons. First, when the cognitive aspects of perception are fixed, we are assured that perceivers will be restricted in what they perceive and can't perceive just what they wish. But that doesn't mean that what they perceive will be accurate. A good example of this potential misperception is the Müller–Lyer Illusion, which is often taken as a classic example of modular processing. Here, one has two straight lines that are in fact of equal length but that (relative to whether they depict inward or outward pointing arrows) appear to have different lengths. This illusion is taken as supportive of the modularity of perception, since one can't dispel the illusion by thinking—even when one learns that the lines are of equal length, one nevertheless sees one as longer than the other. But the example also illustrates the problem we are citing here, that modularity has no necessary connection to reliability—not being able to cognitively penetrate one's perceptual processes is no guarantee that these processes will be more truthful. To be sure, proponents of modular perception have the option of citing the evolutionary history of animals to support the thesis of modularity—for example, as Cosmides and Tooby (1994) suggest, encapsulated perceptual processes work faster than general purpose cognitive processes and so have survival value—and from here might argue on behalf of the reliability of modular perception on the grounds that animals need reliable perceptual processing to survive. Leaving aside the question of whether modular perceptual processes are actually selected because of their reliability, it's surely the case that our local environment is now so different from that of our evolutionary past that modular perception can no longer be firmly trusted, as we found with the Müller–Lyer Illusion.

It's worthwhile considering too that in scientific research observational processes are greatly enhanced by artificial means and that whatever fixedness is contained in modular perception can be straightforwardly overwritten by technological enhancement. In other words, perception that would otherwise be cognitively impenetrable is penetrated all the same by the prosthetic inventions of experimenters. Such an invention can be as simple as using a ruler to measure the true length of the lines in the Müller–Lyer Illusion. So theory, by the circuitous route of experimental design, can inform observational results in a way that goes beyond perceptual modularity, and once more we are left with the challenge of theory dependence for the objectivity of observation that prompts philosophers such as Kosso to adopt the strategy of independence of an account.

My assertion, nevertheless, is that the hazard of theory dependence is not as great as a proponent of independence of an account seems to think. There are a number of reasons for this. One is the fact that, despite the epistemic nature of observation, all observation is built from a basis that is nonepistemic. We all observe things nonepistemically, to begin with, and as we mature intellectually we begin interpreting what we observe, first through our experience of using language and then more rigorously by the development of theories. In other words, we never lose the nonepistemic aspect of our observations, and for this reason the imposition of our theoretical preconceptions on what we observe is never as complete and thorough as the proponents of independence of an account seem to fear. Let me illustrate this point by alluding to the work of a philosopher who has, surprisingly, made the prospect of circular theory-dependent observation an ongoing concern for philosophy. Here I am referring to the work of Thomas Kuhn.

Kuhn, in *The Structure of Scientific Revolutions* (1996), is famous for emphasizing the conservative nature of scientific inquiry, the feature of scientific inquiry in which scientists are trained to observe the world in preset ways organized by the paradigmatic assumptions of the scientific tradition that informs their instruction. I would suggest that some of the enthusiasm behind the restriction to independence of an account in the literature is in an important sense a result of efforts to undermine the Kuhnian view of normal science. Kuhn emphasized the theory-dependent nature of scientific observation and the resiliency of scientists to utilizing alternate ways of examining the natural world—alternate, that is, to the paradigm under which they are working. Nevertheless, some comments made by Kuhn in *Structure* confirm my optimistic claim above that, despite the epistemic character of observation, there is no particular concern that the theory motivating an observation will irrevocably lead to results that confirm this theory. Here I am alluding to Kuhn's reference to the psychological experiments performed by Bruner and Postman on the perception of anomalies (see Kuhn 1996, 62–65). The subjects in those experiments are shown anomalous playing cards, such as a red six of spades and a black four of hearts, and in the usual case (with relatively short exposure times), the subjects categorize the cards in a nonanomalous fashion (e.g., a black four of hearts was identified as either a black four of clubs or spades). So let us imagine for a moment that these subjects are testing the hypothesis 'this deck of cards is a standard deck'. What proponents of independence of an account worry will happen is that observations will be found to confirm this hypothesis precisely because this hypothesis is informing their observations: The subjects anticipate a normal deck of cards and are led to 'see' a normal deck of cards, and so the hypothesis 'this deck of cards is a standard deck' is confirmed.

However, something very surprising subsequently occurs in the experiment. As the subjects are exposed to the anomalous cards for increasing amounts of times, they start to become aware of differences in the cards. To some these differences become obvious fairly quickly. To others, the switch is more protracted and painful. The point is, theory ladenness notwithstanding, observations indicating a reality not conforming to one's theoretical expectations have a way of intruding on each person's psychic life. The involuntariness of this intrusion, its causal nature, is for me one of the more lasting images of the Bruner and Postman experiment and casts doubt on the need to adopt independence of an account to assure the objectivity of observations. Even when people assiduously preconceptualize their observable world in a certain way, it can still become impossible for them to see it that way, if the world isn't that way.

The phenomenon we are describing here, that people do not necessarily observe what they anticipate observing, is due to the nonepistemic character of observation. Put in causal terms, what we observe is due in part to those features of the world that cause our observations. At all times, what we describe ourselves as observing is a product of the conceptual framework we use to comprehend these observations. For instance, the subjects in the Bruner–Postman experiment are able to control what is referred to by the terms 'heart' or 'spade' or 'four'. But once the referents of these terms are fixed, it is, as we have seen, not up to the subjects to describe their observations however they like. That is, what they observe will be filtered through this referential framework, yet this framework will not determine what they observe. For this reason, the spectre of observing what we theoretically conjecture we will observe is not as grave as some fear.

However, one need not draw on the nonepistemic or causal nature of observation to defuse a concern with observational circularity. In particular, in a case where there is an overriding concern that theoretical anticipations will play a determining role in guiding observation, there is no particular epistemic merit to be derived in adopting independence of an account. The sort of case I have in mind, one in which there is a strong propensity for observation to be influenced by background theory, occurs frequently in social scientific research and clinical medicine. To take a simple example from medicine, suppose a drug increases an ill patient's average life-span from three months to six. Are we observing an improvement of health? Answering this question depends a great deal on one's particular view of health, and depending on this view one will see an improvement in health or not. Similar examples can be drawn from the social sciences. When we see someone getting her nose pierced, are we witnessing deviant behavior? Or, if we see someone shuffle his feet in a peculiar way in a social setting, are we watching body language? How one responds in each of these cases no doubt strongly depends on one's prior convictions regarding what counts as 'deviance' and 'body language'.

As a result of the propensity of observations in the medical and social sciences to be heavily influenced by the preconceptions of researchers, it is common practice in these fields to require that experimentation be performed using double-blind tests. That is, when a researcher is experimenting with human subjects, it is recommended that certain important facts be withheld both from the subjects participating in the study (a single-blind test) and from the experimenters (a double-blind test). To illustrate, suppose a medical scientist is testing the effectiveness of a drug by comparing the effects of this drug on those receiving it with those receiving only placebos. With double-blind testing, we ensure that the experimenter is unaware of who receives the genuine drug and who does not. We do this because an overzealous researcher, believing the drug to be beneficial and keen to see improvements in patient health, may 'see' these improvements even if they are not present. It is by concealing from the researcher the facts about who is receiving the drug that we block the influence of the researcher's expectations on the results.

Now one might think that the sort of interpretive biases we have been describing, as dealt with through double-blind testing, could otherwise be resolved by the adoption of independence of an account. To make the situation clearer, suppose we have an experimenter who is testing a theory T, which, let's assume, states that a drug has beneficial effects on patient health. The problem is that T, when believed, impels the experimenter to 'see' improvements in patient health when the drug is taken. The double-blind condition, in turn, removes from the awareness of the experimenter the information about which patient is taking the drug. Such a condition, it is claimed, effectively inhibits 'reading one's theory into' observations. But, under the circumstances of double-blind testing, it won't matter whether independence of an account is satisfied (or not). The experimenter, we can suppose, still has her preferred theory T in mind, believes it as ever, and is intent on retrieving observations that confirm her theory, yet with a double-blind test the influence of her theoretical preconceptions is all for naught. With the imposition of double-blindedness as a methodological constraint independence of an account is a nonissue, and so its violation is acceptable.

Let us then consider a case where the double-blindedness condition is not imposed, where reading one's theory into observations poses a hazard that we need to worry about. Where we lack a double-blindedness condition, is there reason to adopt independence of an account? To begin with, it is true that without the double-blindedness condition researchers in some situations will see what they want to see in the data. In particular, this will occur in situations where the determination of an observational result is highly interpretive, such as we have found in experiments conducted in the medical and social sciences. And, to be sure, these more interpretive experimental situations are not at all ideal. But we need to emphasize that these sorts of situations are problematic whether or not the theoretical preconceptions informing our observations are those that are under test. That is, our problems here are not particularly bad just in the case where our observations are laden with the theory under test. It is, more basically, the ladenness of observations by whatever theory the observer has in mind, under test or not, that is a concern. The problem here is the highly interpretive, highly flexible nature of the observations, their theoretical malleability. This is the source of the potentially misleading character of these observations, leaving aside the issue of whether it is the theory under test that is informing these observations.

As such I believe we are left with the following conclusion to draw as regards observations made in the medical and social sciences and the need for double-blind tests. If because of the highly subjective and interpretive nature of our observations we decide to use double-blind tests, it follows that independence of an account is not a needed requirement. Alternatively, where we do not have recourse to double-blind tests, the problem we find with the data has nothing to do with the fact that the data is interpreted in accordance with one's preconceptions, where these preconceptions are themselves under test, but with the unreliable nature of the data itself, regardless of what theory is under test. That is, adopting independence of an account in no way improves the situation since the situation is so rife with interpretive bias that, if it is not the theory under test that is informing the observations, then it is some other, perhaps even more ill-chosen theory that is doing the informing. Here we might imagine a case where a social science researcher is performing observations to see if a certain kind of music promotes deviant behavior and, being a believer in the view that it does, starts to witness deviant behavior occurring under the influence of such music. The critic, a proponent of independence of an account, might object that such theory testing is unreliable because of the influence of the researcher's beliefs. But then the critic, when pressed to find a better approach in identifying instances of deviance, might only be able to provide theoretical strategies that are themselves highly interpretive and unreliable, and so the situation is not improved from the case where observations are interpreted in accordance with the theory under test. Indeed, the theory under test might be the best theoretical perspective with which to interpret the observations—that might be why it is the theory under test. In such a case, by removing the influence of this theory by means of double-blind tests, one would be reducing the reliability of the observations. For these reasons, I do not see what value there is to adopting independence of an account even with respect to the highly interpretive observations found in the medical and social sciences.

In further exploring whether there is merit in the independence of an account requirement, it is worthwhile to distinguish two ways in which one's theoretical preconceptions can influence observational results. The first way is this: Suppose after retrieving certain observational results a scientist is in a position to adjudicate whether these results support her theoretical perspective. We might have a problematic circularity here if she endorses the confirmatory significance of these results only if they conform to her theoretical perspective and negatively assesses the results otherwise. As such, there is motivation to prohibit such a possibility, motivation that we might entrench in terms of a methodological dictum: 'in testing a theory using an observation, do not use that theory in evaluating the evidential significance of the observation'. As I will put it, this is to portray independence of an account as an 'evaluative' principle—it recommends avoiding the use of the theory under test in evaluating the significance of observation for this theory. Alternatively, one can use the theory under test to generate observational results, such as using it in the design of an experimental apparatus, in providing guidance on how an apparatus should be handled, in processing raw data into a usable format, and so on. This alternate use of theoretical assumptions in the deployment of observations I call its 'generative' use, and one can accordingly understand independence of an account as a 'generative' principle. This principle we can formulate as follows: 'In testing a theory using an observation, do not use that theory in generating the observation'.

The question now facing us is whether the epistemic status of independence of an account differs depending on whether it is read evaluatively or generatively, and I wish to argue that it is a dispensable principle in both senses, though for different reasons. Let us first consider the generative approach.

To help us in assessing independence of an account as a generative principle, I focus on the work of Martin Carrier and his discussion of work by Joseph Sneed. Carrier (1989) discusses the epistemic problems raised by so-called Sneed-theoretical terms, terms for which '*all* means of determining the truth-value of statements involving the [term] presuppose the truth of the laws of the theory in question' (411; quoted from Sneed 1979, XVIII; Sneed's italics). Carrier's point is that some observation statements (e.g., 'this object has mass *m*') when used in testing certain theories ('Newton's second law') are Sneed-theoretical (here, 'presuppose Newton's second law') and so cannot properly function in testing these theories. As a result, he offers the recommendation that we avoid observational descriptions that use Sneed-theoretical terms, which from our perspective is precisely the recommendation to adopt the principle of independence of an account. Moreover, as Carrier (1989) makes clear, he is interpreting this principle generatively, as revealed by his comments,

> Sneed considers mass and force to be [Sneed-]theoretical relative to classical particle mechanics which he sees characterized by Newton's second law, i.e., the equation of motion ($F{=}ma$). This entails that all procedures for measuring masses and forces should make use of [the] second law. (412)

As such, Carrier's concern that motivates his adoption of independence of an account seems to be that it becomes impossible to test Newton's second law in using mass and force observations since we use Newton's law to generate these mass and force observations to begin with. Carrier's counsel is to avert this troublesome circularity by utilizing mass and force observations that can be generated without the use of Newton's second law.Do we have a troublesome circularity here? To make these issues more concrete, suppose we observe an object to have mass *m*, where generating this observation involves a procedure utilizing Newton's second law. And let us imagine that, given certain other theoretical commitments of ours, commitments differing from our commitment to Newton's second law, we expect an object with mass *m* to have a particular observable effect—for example, we anticipate that this object would require a significant amount of physical effort if we were to lift it. Finally let us suppose that, to our surprise, this other observable effect does not come about—the object in fact is quite light. What should our conclusion be? We have a number of options, one of which is the following. We can argue that the procedure by which we generate the observational result, 'the object has mass *m*'—a procedure that uses Newton's second law—is flawed because, in this sort of case, Newton's law is false. Of course, Newton's second law is highly entrenched, and it is doubtful that we would challenge it based on how confident we are in our kinesthetic sensations of force. But the point, in any case, is that there is no logical obstacle to such a challenge, based on the supposed circularity of the testing process. Whether we accept the challenge will depend on how committed we are to Newton's second law. If this commitment is fundamental to us (as it actually is), then we will resist the refutation. But there is nothing in having presupposed Newton's second law in the generation of observations that guarantees its protection from being falsified by these very same observations, should these observations conflict with other observations (or indeed with other theoretical commitments).

It is worthwhile generalizing the above argument, for nothing hinges on the details of the specific case study involving Newton's second law. The result we have drawn is intimately connected with a broader feature of experimental testing, its Duhem–Quine nature. That is, when there are untoward experimental results, the issue always arises concerning where to pin the blame, and one always has the option of questioning those theories underlying the observational or experimental process—even in the case where these theories are the very ones being tested. Consequently, where one's observations are informed by a theory T, so long as an experimentalist is willing to question T given untoward results, there is no need to worry about vicious circularity in one's testing procedure. The crux, we might say, is the experimenter's attitude toward the testing situation, whether she is genuinely open to questioning the theory informing the observations. If she is, then she is free to adopt that side of the Duhem–Quine divide that questions the theory underlying the observations.

My overall conclusion, then, as regards the 'generative' version of independence of an account is this: I argue that we need not be preoccupied with generating observations that do not 'presuppose' the theory under test. We should generate observations, rather, with the goal of producing reliable observations, however one conceives this should be accomplished and even if this involves assuming the theory under test in generating these observations. It will not necessarily follow, in any event, that the theory informing these observations is uniquely privileged.

What then can we say about the evaluative version of independence of an account, that is, 'in testing a theory using an observation, do not use that theory in evaluating the evidential significance of this observation'? My claim is that, where there is an emphasis on the empirical evaluation of theories, one need not be concerned about violations of independence of an account, so understood. My reasoning is as follows. Suppose we have a case where a scientist generates observations and then compares these observations to the predictions of some theoretical hypothesis. Suppose further that the observations falsify these predictions and that the scientist, in an effort to salvage the hypothesis, constructs a scenario where he explains away the deviant results. There are, he perhaps explains, certain abnormalities in the testing situation that render the observations irrelevant; or maybe his hypothesis, he argues, is not really meant to cover the sorts of situations described in the experiment; or again, he notes that with a slight revision to his theory (which he says should have been there to begin with) the theory does make the correct predictions. There are obviously a number of options for the scientist to pursue here, and they have the appearance of the sort of ad hoc, circular revisions proscribed by the evaluative version of independence of an account. He seems to be using his adherence to the theory under test to adjudicate his judgment about the evidential significance of observations. However, there is no reason for us to conclude necessarily that these are flawed revisions, for they may be motivated empirically. For example, the scientist may claim that the cited abnormalities in the testing situation can in fact be observed, or that there is a worthwhile, empirical justification for the claim that the hypothesis was not intended to cover the experimental situation under consideration, or even that the suggested revision to his hypothesis that allows him to capture previously anomalous data is empirically based. In general, when a scientist reconceives his theory to accommodate a negative observed result, the empirical motivation for this reconception has to be studied before we draw any definite conclusions about the acceptability of the scientist's manipulations.

My belief is these ancillary empirical questions can have precisely the effect of restoring the objectivity of the testing process, for there is no guarantee that they will turn out the way the theorist hopes or expects. Empirical investigation may not reveal any abnormality in the apparatus; the projected revision to the hypothesis may not be empirically sustainable; and so on.

To be sure, the scientist could repeat the process, again making revisions to his broader theoretical perspective in an attempt to reconcile his particular hypothesis with untoward results, and again flouting the evaluative version of independence of an account. But this may only put off the inevitable if there are even more disruptive empirical consequences to follow. Or possibly, as the case may be, these may be exactly the sorts of moves that are needed to restore and establish the scientist's theoretical perspective in an improved form. However these revisions go, the overall point to be emphasized is that it is wrong to say that objectivity is necessarily compromised in violating the evaluative version of independence of an account. So long as a scientist retains an empirical sensitivity and aspires to put to empirical test any revised understanding of the world, there is no worry that he will perpetually and consistently maintain his favored theoretical hypothesis.

Finally, I would go further and claim that the dictum 'in testing a theory using an observation, do not use that theory in evaluating the evidential significance of this observation' is altogether unwise advice if it actively dissuades scientists from maintaining their hypotheses in the face of contravening evidence. As Kuhn notes, it is the normal course of affairs for scientists to maintain their theories in the face of contrary evidence and to reinterpret this evidence accordingly. It is the character of theories that they live in a 'sea of falsifications'. Ultimately, in this context, the legitimacy of such reinterpretations is a matter of degree: The more evidence needs to be reinterpreted, the more empirical pressure there is to change theories. Eventually it can happen that the empirical pressure becomes enough to force a change, at which point it would be a mistake to continue reinterpreting experimental situations in accordance with one's theoretical preconceptions. But the mistake here would not be the mistake of having violated the evaluative version of independence of an account—the reinterpretations all along were such 'violations'. The mistake would be one of ignoring a growing preponderance of negative evidence and insisting, nevertheless, on one's theoretical perspective.

SUMMARY

In this chapter, we have examined three different approaches to defending the epistemic significance of robustness reasoning: (a) a probabilistic approach, (b) a pragmatic approach and (c) an epistemic independence approach. My criticism of these three approaches notwithstanding, one can nevertheless identify a core argument for robustness (ultimately deriving from the no-miracles argument for robustness) that is, in all likelihood, the ultimate source of the support robustness reasoning enjoys. In chapter 6 we return to an assessment of this core argument. In the interim, chapters 2 to 5, we examine a variety of scientific case studies that reveal the true value of robustness reasoning for scientists (not very much) and that provide insight into how scientists actually go about establishing the reliability of observed results.

Chapter 2

The Mesosome: A Case of Mistaken Observation

In the preceding chapter we examined various philosophical approaches to defending robustness reasoning. In the next four chapters, we will consider the question of robustness from a historical perspective. The idea, generally speaking, is to see if robustness reasoning is in fact used by practicing scientists. If not, this is a result that would have key importance for the philosophical situation regarding robustness. In such a case, philosophers who are supporters of robustness would have to either contest the details of the historical case studies, suggest that the choice of case studies is biased, or more drastically claim that the participant scientists were unaware of, even confused about, the value of robustness.

In order to address the question of whether our choice of case studies is biased, I examine in this chapter a case study that at least one philosopher argues is a clear illustration of how scientists use robustness reasoning. The case concerns the purported discovery of the bacterial mesosome that Sylvia Culp argues involves the application of robustness reasoning (Culp 1994, 1995), and we delve into this case to see whether she is correct. Further, in chapter 4, we investigate what is perhaps for philosophers the most celebrated case of robustness reasoning: Jean Perrin's argument for the reality of atoms. On the view of many philosophers (such as Cartwright 1983, Salmon 1984, Kosso 1989, and, more recently, Stegenga 2009), Perrin's reasoning is a paradigmatic example of how a scientist has effectively used robustness reasoning to defend an experimental conclusion. In chapters 3 and 5, we explore some recent astrophysical research that provides some novel test cases for robustness. Chapter 3 examines the supposed reality of weakly interacting massive particles (WIMPs), a candidate subatomic particle held by some to constitute cosmological dark matter. Chapter 5 investigates recent empirical arguments for the reality of dark matter itself, as well as arguments for a different astrophysical phenomenon, dark energy. The astrophysical cases are chosen primarily for their broad interest and for the fundamental nature of the research: Many scientists (and laypeople) are interested in this work, and the research promises to inform our deepest understanding of the nature of the physical universe.

Before we engage these historical case studies, a brief word is due regarding the sense of robustness we will be working with. Essentially, the argument for robustness that has survived our analysis of chapter 1 is the 'core argument' that purportedly isolates an epistemic advantage to robustness reasoning. On this argument, robustness reasoning involves the deployment of independent physical processes that converge on a particular observed result. Culp for her part interprets the independence intrinsic to robustness as epistemic independence—that is, observational processes are independent in that the theoretical assumptions that underpin these procedures are different—and we suggested in the previous chapter that interpreting independence in this way fails to account for the presumed informative value of robustness reasoning. Nevertheless, reasoning on the basis of epistemic independence could generate the advantages found in the core argument in that observers, when working with different theoretical assumptions, thereby also utilize different physical processes (as is required in the core argument). In general, the goal of robustness reasoning in all the historical cases we examine in this book is to generate observed reports that have an increased likelihood of truth, as opposed to results that have particular pragmatic virtues (as with the Wimsattian approach). The virtue of the core argument is that it actually makes a case for why this goal is achievable using robustness reasoning.

With respect to our first case, the case of the bacterial mesosome, it is Sylvia Culp's contention that experimental microbiologists, after initially maintaining that mesosomes were real components of bacteria, subsequently learned that mesosomes were artifacts after making concerted use of robustness reasoning. Thus, for her, the bacterial mesosome forms a successful 'test-case' (which is her expression) for the applicability of robustness reasoning. I will argue, on the contrary, that with a closer reading of the mesosome episode, it becomes apparent that robustness reasoning was not at all the epistemic strategy scientists used to reveal the false reality of mesosomes. As I strive to show, scientists during this episode use a different form of reasoning, which I call 'reliable process reasoning'. By such a form of reasoning I mean nothing more complicated than, first, identifying a process that has the character of producing true reports with inputs of a certain kind, and second, recording that one actually has an input of this kind. Of course what is left out in describing reasoning along these lines is a description of why a process is deemed reliable. As I illustrate below, this judgment often rests on the grounds that the process avoids a characteristic sort of error. But sometimes the reliability of a process is simply 'black-boxed', and the sort of argument that uses reliable process reasoning will follow the simplistic schema just outlined. I regard this feature of how experimentalists argue in the mesosome case to be significant and to exhibit a very different kind of thinking than robustness reasoning. It's the difference between asserting that one is observing something correctly because one's observational process is (inherently) reliable, as opposed to asserting that one's correct observation is justified by the convergence of the output of one's observational process with the outputs of different, observational processes. In the context of reliable process reasoning, it's still possible to provide support for the claim that a process is reliable, and below we see examples of experimentalists doing just that. Often this amounts to a demonstration that the process evades certain critical errors. We don't find, in any event, robustness reasoning being used for the purposes of this task.

We turn now to examining the mesosome case. The discussion of this case was initiated by Nicolas Rasmussen (Rasmussen 1993). Rasmussen's take on the episode is sociological in the sense of the strong programme; that is, he doubts that the mesosome episode was rationally resolved in the way many philosophers of science would prefer to think of it. For him, various nonepistemic, social forces were in play that culminated in the mesosome being relegated to an artifact. It is in response to Rasmussen's antirationalism that Culp sets forth her robustness interpretation of the episode. In objecting to Culp's robustness approach, I don't mean to abandon her agenda of restoring the epistemic credentials of the episode—it's just that I think she took the wrong tack in going the robustness route. At the end of this chapter I take up and rebut Rasmussen's sociological (strong programme) interpretation of this experimental work

INTRODUCING THE MESOSOME: RASMUSSEN AND CULP

When the electron microscope was used in the middle of the 20th century to examine the ultrastructure of bacteria, there was a surprising revelation. It had traditionally been thought that bacteria were organelle-less: They contained no mitochondria, ribosomes, Golgi apparatus and so on. Then, electron microscopic work performed by George Chapman and James Hillier (Chapman and Hiller 1953) revealed what was apparently a bacterial organelle, one they initially called a 'peripheral body' but that later became known as the 'mesosome' (Rasmussen 1993, 233–234). Pictures of mesosomes were produced by electron microscopists from the 1950s through to the mid-1970s, with hundreds of papers appearing in prestigious journals containing experimental results describing mesosomic structure, function and biochemistry. After 1975, however, the views of the microbiological community changed: Mesosomes were no longer asserted to be bacterial organelles but rather claimed to be artifacts of the process by which bacteria are prepared for electron-microscopic investigation, a view that persists to the present day.

The mesosome episode is a fascinating one from the perspective of scientific rationality because it shows how contemporary scientists (like fallible humans everywhere) can be drawn on rational grounds to believe a claim and later be equally drawn on rational grounds to reject it. Nicolas Rasmussen, for his part, derives a somewhat negative conclusion from this episode regarding the rationality of science:

> It will emerge that although the long view of philosophy might take [certain] epistemological principles as constant landmarks, in actual scientific practice, epistemology is in flux on all the less abstract levels: the proper formulation of a criterion, what tactics

> properly apply it, which criteria are most important, and which tactics among many instantiating a given criterion are best—all are constantly open to negotiation. The turmoil of actual science below the most general level of epistemological principle casts doubts upon efforts in the philosophy of science to produce validation at that level. (231)

Specifically, Rasmussen finds questionable the role of robustness in the mesosome episode:

> I show that independent theory of methods and instruments is not in practice depended on by biological electron microscopists to assure reliability of observations, or to decide reliably between conflicting observations. (231)

For Rasmussen, this is not to say that bacterial microscopists (and scientists generally) do not use robustness reasoning. He thinks they do (Rasmussen 2001, 642) but that such reasoning (along with the other principles of reasoning philosophers are prone to suggest) is too abstract, works at 'too low a level of resolution' (as he puts it), to effectively adjudicate scientific controversies. His view echoes a familiar refrain from sociologists of scientific knowledge such as David Bloor, Barry Barnes, Harry Collins and many others who find abstract philosophic principles to be of limited use in understanding scientific practice and who suggest, then, that to formulate a more complete view of scientific work one needs include nonepistemic factors such as 'interests' (Rasmussen 2001, 642), 'intuition, bias due to training and a host of other personal and social factors traditionally regarded as external to science' (1993, 263).

Rasmussen's challenge to philosophers was taken up by Sylvia Culp who argues (Culp 1994, 1995) that Rasmussen's history of the mesosome episode is incomplete. As Culp suggests,

> A more complete reading of the literature shows that the mesosome 'ended up an artifact after some fifteen years as a fact' [quoting Rasmussen] because the body of data indicating that bacterial cells do not contain mesosomes was more robust than the body of data

> indicating that they do. Mesosomes were not consistently observed when electron microscopists attempted to observe mesosomes both by varying conditions with already established sample preparation techniques and by using newly developed sample preparation techniques. (1994, 47)

In other words, on her view, the principle of robustness is not too vague, nor too abstract, to effectively serve the role of deciding on this scientific controversy (and, by extension, other controversies). As such, her paper (Culp 1994) contains a detailed examination of various experiments that she thinks demonstrates how, by using robustness, microbiologists became assured of the artifactuality of the mesosome. From my perspective, I am uncertain whether Culp's detailed examination is detailed enough, and below I describe a number of the relevant experiments with the aim of showing that robustness reasoning was not used by microbiologists in demonstrating the artifactuality of mesosomes. But before I begin that description, there are various features of Culp's approach that we need to address. First, she regards the robustness reasoning scientists are using as leading to a negative result—as showing that mesosomes do not exist. My sense is that this is a risky form of robustness reasoning. Consider that the sum total of all observations prior to the invention of the electron microscope never revealed mesosomes—and without a doubt the majority of these observations were independent of one another. Still, such a vast convergence of independent results goes nowhere in showing that mesosomes do not exist for the simple fact that none of the underlying observational procedures had any chance of revealing the existence of mesosomes, if they were to exist. In other words, there is a need here for a sort of minimal reliability requirement such as we described in chapter 1 with reference to Sober's argument for robustness. We let proposition P stand for 'mesosomes don't exist', $W_i(P)$ stand for 'witness W_i asserts that P' and, accordingly, require that

$$(S) \quad P[W_i(P)/P] > P[W_i(P)/-P], \text{ for } i = 1,2,\ldots$$

It follows that, if the observational procedures we are using are so bad that they would never reveal mesosomes even if they existed (i.e.,

$P[W_i(P)/-P] \approx 1$), then the fact that mesosomes don't appear is not proof that mesosomes don't exist, even if the negative results are robust. My point is that when we are engaged in highly speculative research (as with the experimental search for mesosomes) in which the reliability of observational procedures in detecting a unique entity is subject to doubt, the occurrence of 'negative' robustness where we confirm the nonexistence of this entity by a variety of observational methods does not tell us much. This is true despite the fact that, in the majority of cases, we are indeed able to reliably track the nonexistence of the sought-for entity—for example, we have great success in tracking the nonexistence of mesosomes in environments barren of bacteria.

The second feature of Culp's approach we need to appreciate is the sense in which, for her, observational procedures are independent. She is concerned with what she calls 'data-technique circles'—cases in which one's theoretical assumptions (incorporated in one's observational 'technique') strongly influence how raw observational data are interpreted and, accordingly, what interpreted observational data are produced. Following Kosso, she advocates the need for independence of an account (though she doesn't use that terminology), arguing that 'it is possible to break data-technique circles by eliminating dependence on at least some and possibly all shared theoretical presuppositions' (1995, 441). Similar to Kosso, the path to eliminating such dependence is by using multiple experimental techniques that converge in their results: 'This dependence can be eliminated by using a number of techniques, each of which is theory-dependent in a different way, to produce a robust body of data' (441). Of course, as we have raised doubts about independence of an account, the need for robustness as Culp sees it is also subject to doubt. But here our concern is solely historical: Do the participant scientists in the mesosome episode utilize robustness reasoning in arguing against (or perhaps for) the reality of mesosomes, as Culp suggests? If so, this is reason to think that robustness has a place in the philosophical repository of epistemically valid tools for ensuring the accuracy of observational procedures.

Very briefly, Culp asserts that a number of experiments performed by microbiologists from 1968 to 1985 show the following: For the set of techniques that could be used to reject the mesosome, there is a higher degree of independence among the theories used to interpret electron micrographs than for the set of techniques that could be used to support the mesosome (i.e., members of this latter set all depend on theories about the effects of chemical fixation or cryoprotectants; 1994, 53). To assess whether Culp is correct in this assertion, I look closely at the experiments she examines, as well as some further ones. In due course it will become clear that robustness does not play the fundamental role that Culp ascribes to it in her understanding of this episode. In a limited sense, then, I agree with Rasmussen's denial of the pivotal role of robustness. Rasmussen and I part ways, however, when it comes to assessing why the mesosome was subsequently relegated to the status of an artifact. For me, as we shall see, it was a substantial epistemic matter and not a matter of social, political or other nonepistemic interests.

THE MESOSOME EXPERIMENTS

There were a number of the microbiological experiments performed between 1968 and 1985 dealing with the existence of mesosomes, and, for the most part, we will be considering the same experiments discussed by Culp (1994). Let's start by making some comments about what mesosomes look like and where they are found. Mesosomes occur in bacteria as enclosed membranous structures and are seen sometimes as empty sacs, sometimes as sacs within sacs (vesicular mesosomes) and sometimes as stacks of membranes (lamellar mesosomes). Occasionally they are near the center of a bacterium (near the nucleoid, where one finds a bacterium's DNA); other times they are near the periphery of a bacterium, that is, near the plasma membrane. Sometimes bacteria contain many mesosomes and sometimes only one or two. In a collection of observed bacteria, many, some or none might contain mesosomes. In addition, mesosomes can range in size from small to large. With such dramatic variability in mesosome frequency, size, shape and so on, one needs to make some assumptions about when it is true to say that bacteria have been observed to contain mesosomes. Here I follow the practice of most experimental microbiologists who have worked on mesosomes by asserting the presence of mesosomes whether or not they were observed to be big or small; central or peripheral; empty sacs, vesicular or lamellar. I also adopt no preconception regarding how many mesosomes one should expect to see in bacteria or about what proportion of visible bacteria should contain them, leaving these judgments to the experimenters themselves in their assessments.

It is worthwhile pointing out that to prepare bacteria for electron microscopic investigation, we must manipulate them in certain ways to withstand the harsh environment created by an electron beam. Again, the following is a simplification, but it is in any event a simplification used by Culp. (Note that the following discussion concerns the state of technology during the time period at issue.) There are four ways in which bacteria are manipulated to prepare them for the electron microscope: (a) prefixed, (b) fixed, (c) cryoprotected and/or (d) sectioned. Prefixing and fixing might involve turning the bacterium into a piece of plastic; that is, bacteria are 'polymerized', making them much easier to section (i.e., cut). Typical chemical reagents used at this stage are osmium tetroxide (OsO_4) and glutaraldehyde (GA). Cryoprotection is used when the preparative process involves freezing bacteria; cryoprotection is needed to hinder the formation of ice crystals in the bacterium, for such crystals, presumably, could alter the morphology of a bacterium. Sectioning involves either cutting a bacterium into two-dimensional planes—much like cutting a very thin disk out of a tree's trunk—or coating a frozen, cut bacterium with a metal and dissolving away the organic matter, leaving behind a metallic replica that mimics the contours of a bacterium's internal structure. This latter procedure does not sound much like sectioning, yet Culp lists it under this rubric and so we will follow her on this matter for the sake of continuity. Now, with the above manipulations—prefixing, fixing, cryoprotection and sectioning—there are innumerable variations, more than we have the space to consider. I address them below as the need arises.

Some key experiments were performed by Remsen (1968), who found mesosomes by freeze-etching with no prefixing, no fixing and no cryoprotection, and Nanninga (1968), who observed mesosomes by a similar regimen, except he used a cryoprotectant; he found mesosomes whether cryoprotection involved glycerol and sucrose, or alternatively, glycerol and no sucrose. Nanninga (1968) also observed mesosomes with freeze-etching and with thin-sectioning, where GA was used as a prefixative, OsO_4 was used as a fixative, and there was no cryoprotection. We find then with this limited, initial set of experiments that the relevant techniques have been varied in a significant number of ways (with no doubt correlative changes in what theoretical assumptions are needed), and the same result is occurring. Whether or not GA is used as a prefixative, mesosomes are seen. Whether or not glycerol (with or without sucrose) is used as a cryoprotectant, mesosomes are seen. Whether or not thin-sectioning or freeze-etching is used, mesosomes are seen. So far, robustness is telling us that mesosomes exist.

This pattern of finding 'robust' experimental support for mesosomes continued into the 1970s and early 1980s. Silva (1971) explores the use of thin-sectioning. Mesosomes were observed on this approach when no cryoprotection was used, OsO_4 was used as a fixative, and whether or not prefixation involved OsO_4 and calcium or OsO_4 and no calcium. On the other hand, when the OsO_4 prefixation step was omitted, Silva reports that 'simple and usually small intrusions of the cytoplasmic membrane were found' (230). Silva declines to call these membranous intrusions 'mesosomes', and, in summarizing his results, he comments, 'When prefixation was omitted, mesosomes were not observed' (229–230). Culp, too, in presenting Silva's results, lists the no OsO_4 case as a nonobservation of mesosomes; following Silva (1971), she counts as a mesosome only something that is large and centralized (see Culp 1994, 51, Table 3). However, Silva's disinclination to call these small membranous intrusions mesosomes is atypical. Microbiologists at that time, and currently, are prepared to call these smaller bodies mesosomes, and, in fact, Silva himself calls them mesosomes in later work (Silva et al. 1976). I suggest, then, that we count Silva's observations of small, membranous intrusions, where OsO_4 is omitted as a prefixative, as observations of mesosomes. Consequently, it appears that robustness is again supportive of the existence of mesosomes.

The results from Fooke-Achterrath et al. (1974) are less decisive, but, as Fooke-Achterrath et al. interpret them, they are supportive of the claim that mesosomes exist. When bacteria were prepared at a lower temperature than usual (4°C), prefixed with a variety of different concentrations of OsO_4 (.01%, .1%, .5%, 1% and 3.5%), fixed at 1% OsO_4 and thin-sectioned, small, peripherally located mesosomes were found in 10% to 20% of the observed bacteria. Also, whether or not glycerol is used as a cryoprotectant, freeze-etched cells (again prepared at 4°C) revealed mesosomes 15% of the time. Though one might find these results to be inconclusive, Fooke-Achterrath et al. take them to provide positive support for the existence of small, peripheral mesosomes. As they say, 'The number of [small, peripheral] or "true" mesosomes per cell is 1 or 2 and does not fluctuate' (1974, 282). On the other hand, bacteria prepared at 37°C, prefixed with either .01%, 1%, .5%, 1% or 3.5% OsO_4, fixed at 1% OsO_4 and thin-sectioned, exhibited large, centralized mesosomes 50% to 60% of the time. So, if we apply robustness reasoning, we have at worst an inconclusive result and at best a positive result in support of the existence of mesosomes.

Nevertheless, it is worth pointing out that Fooke-Achterrath et al. (1974) express no interest in robustness reasoning as regards their experimental results. Rather, their approach is to assume the greater reliability of freeze-etching techniques. They comment,

> general agreement has been reached that frozen-etched bacteria exhibit a state of preservation closer to life than that achieved by any other method of specimen preparation. (276)

From here, they reason as follows:

> the fine structure of the 'mesosome' in chemically fixed S. aureus specimens represents the structure of the mesosome in vivo only when it corresponds morphologically to its frozen-etched counterpart. Such close-to-life appearance of mesosomes in thin sections was achieved during this investigation only when the specimen was chilled before chemical fixation. (276)

In particular, with low temperature preparations, only small, peripheral or, as they call them, 'true' mesosomes are seen, so the presence of such mesosomes forms the native state of a bacterium. On the other hand, since large, centralized mesosomes are only seen with high temperature preparations, and since these preparations according to them are unreliable, these bodies must be artifactual; they propose renaming them 'technikosomes' (276). As will become apparent, the form of reasoning Fooke-Achterrath et al. are adopting here—justifying observations on the grounds that they are produced by a reliable process—is a common approach with the experimenters we are considering.

Continuing with our catalogue of experiments, Higgins and Daneo-Moore (1974) found mesosomes through freeze-fracturing, whether or not glycerol was used as a cryoprotectant and whether OsO_4 or GA was used as a fixative. They also found mesosomes through thin-sectioning when 1% OsO_4 was used a fixative and when either GA or .1% OsO_4 was used as a prefixative. However, they did not observe mesosomes through freeze-fracturing if no prefixatives, no fixatives and no cryoprotectants were used (whether or not the cells were centrifuged at 5°C or at 37°C, or not centrifuged and poured over ice). A similar negative result was previously found by Nanninga (1971), and reaffirmed by Higgins et al. (1976) and Ebersold et al. (1981): That is, in all these cases, in the absence of prefixatives, fixatives and cryoprotectants, no mesosomes were observed using freeze-fracturing. Again, without prefixatives, fixatives and cryoprotectants, no mesosomes were found by Dubochet et al. (1983) using frozen-hydration, although they did find mesosomes if OsO_4 was used as a fixative. Also, with the freeze-substitution technique, Ebersold et al. (1981) did not observe any mesosomes when GA, uranyl acetate (UA) and OsO_4 were concurrently used as fixatives, nor did Hobot et al. (1985) find any mesosomes (with freeze-substitution) using only OsO_4 as a fixative.

In addition, Higgins et al. (1976) found mesosomes using freeze-fracture methods when GA was used as a fixative and when neither prefixation nor cryoprotection was used. Silva et al. (1976) found mesosomes through thin-sectioning using a variety of OsO_4 concentrations at either the fixative or prefixative stage, as well as when UA was used as a fixative after prior fixation with OsO_4 and GA. No mesosomes were seen, on the other hand, if UA was used as a 'first fixative' with no prefixation (Silva et al. 1976, 103). Silva et al. (1976) also recorded that cells treated with phenethyl alcohol, nitroblue tetrazolium and various anesthetics (tetracain and nupercain) exhibit mesosomes. This pattern of finding mesosomes in bacteria under unusual conditions (e.g., using anesthetics and the like) occurs to this day. Mesosomes are observed in cells treated with haemin, an iron-containing protoporphyrin (Landan et al. 1993), in bacteria exposed to the glycopeptide antibiotics vancomycin and teicoplanin

(Sanyal and Greenwood 1993; see also Santhana et al 2007) and when exposed to the anti-microbialpolypeptide defensin (Shimoda et al. 1995 and Friedrich et al. 2000). Finally, Ebersold et al. (1981) observed mesosomes through thin-sectioning, using GA and OsO_4 as fixatives.

This completes our brief sketch of some of the microbiological experiments investigating mesosomes (see Appendix 4 for a tabular summary). Let us now reflect on these experiments from the perspective of robustness and reconsider Culp's evaluation of the episode. All told, what does robustness tell us? Very likely, if robustness were our chosen experimental strategy, we would be led to support the existence of mesosomes. Usually nonobservations of mesosomes occur under relatively special conditions, that is, in the absence of prefixatives, fixatives and cryoprotectants (Remsen 1968 is a notable exception). Now it seems natural here—given that mesosomes typically appear in the presence of prefixatives, fixatives and cryoprotectants—to suppose that mesosomes are the result of the damaging effect on bacterial morphology caused by such preparative measures. Indeed, this is the story that was subsequently given to explain the occurrence of mesosomes, a story I recite below in presenting the arguments experimentalists use in asserting the artifactuality of mesosomes. But if this is the sort of reasoning experimenters use in disputing the existence of mesosomes, what are we to make of Culp's claim that, with regard to the mesosome episode (her 'test-case' for robustness), the set of techniques that could be used to reject the mesosome was more robust than the set that could be used to support the mesosome? Culp, as an advocate of Kosso's 'theoretical' notion of independence, asserts that there is a higher degree of theoretical independence with those electron micrographs failing to reveal mesosomes than for those micrographs exhibiting mesosomes. This is because, for her, the micrographs containing mesosomes depend on theories about the effects of chemical fixation or cryoprotectants whereas the micrographs without mesosomes do not depend on such theories since they avoid the use of chemical fixation and cryoprotection. But surely, to have this edifying effect, the techniques that generate mesosome-free micrographs *do* depend on theories about the 'effects of chemical fixation or cryoprotectants', in particular, the theory that chemical fixation and cryoprotection damage bacterial morphology and create artifacts. So from Culp's (Kosso-inspired) perspective on robustness, the set of techniques used to reject the existence of mesosomes is no more robust than the set used to support the mesosomes.

Of course, as the history unfolded, mesosomes did come to be viewed as artifacts. So if it wasn't robustness reasoning that played the role in motivating this shift—in considering the data, robustness would lead to either the opposite result or no result at all—why did mesosomes come to be viewed as artifacts? Rather than it being a matter of robustness, I submit that experimental microbiologists were utilizing a different sort of reasoning that I call 'reliable process reasoning', which I now illustrate.

RELIABLE PROCESS REASONING

To start, Silva et al. (1976) in arguing against the reality of mesosomes assert that .1% OsO_4 damages bacterial membranes. They justify this claim by noting that .1% OsO_4 'quickly lyses protoplasts and induces a rapid and extensive efflux of K+ [potassium ions] from B. cereus and S. faecalis [two common bacterial species]' (102). Indeed, they point out, OsO_4 acts in much the same way as known membrane-damaging treatments (e.g., nitroblue tetrazolium). Thus, when cells prefixed with .1% OsO_4 exhibit large, complex mesosomes, one should in fact doubt the reality of these mesosomes since the procedure that generates them is demonstrably unreliable. But why are large mesosomes seen with a lower concentration of OsO_4 and smaller mesosomes seen with a higher concentration? Intuitively, if OsO_4 damages membranes, the situation should be reversed. Here, the feature of OsO_4 as a fixative comes into play. OsO_4 both damages and stabilizes membranes, and at higher concentrations it stabilizes more quickly, thus not allowing as much damage. In this way, Silva et al. are able to explain their observation of large mesosomes in cells prefixed in .1% OsO_4 , and of small mesosomes in cells fixed using 1% OsO_4 or 2.5% GA. On the other hand, (first) fixation with UA leads to the absence of mesosomes, and, as Silva et al. (1976) comment,

> There are good reasons to accept uranyl acetate as an efficient fixative for membranes. Uranyl ions have been shown to have a stabilizing effect action on bacterial membranes and on other

> bio-membranes. Low concentrations of uranyl acetate were found to fix protoplasts. (104)

In other words, fixation with UA is more reliable in that it does not exhibit the membrane-damaging effects found with .1% OsO_4, and since mesosomes are not seen with UA (first) fixation, they must be artifactual.

Silva et al.'s reasoning as exhibited above is an example of what I call 'reliable process reasoning'. Ebersold et al. (1981) argue in a similar fashion against the existence of mesosomes. They first remark that 'traditional methods of electron microscopy such as chemical fixation or freezing in the presence of cryoprotectants are known to induce structural alterations' (Ebersold et al. 1981, 21) for, as they explain, 'Fixatives [and cryoprotected freezing] do not lead to an immediate immobilization of membranes' (21). On their view, the key to preserving (what they call) the 'native state' of a bacterium is to reduce the time needed to immobilize intracellular structures. Unfortunately, Ebersold et al. do not provide much in the way of justifying their belief in the reliability of fast immobilization procedures, except to note that, where specimens are cooled quickly with cryofixation (even without cryoprotectants), ice crystals will not be very large, thus reducing the probability that they will induce structural damage (21). Still, for our purposes, their argumentative strategy is straightforward: They assume the unreliability of slow fixation procedures, and the reliability of fast ones, and then note the conspicuous absence of mesosomes with the latter. In other words, their approach to justifying a no-mesosome result is much like the approaches we saw with Fooke-Achterrath et al. (1974) and Silva et al. (1976)—the testimony of a reliable experimental process is given epistemic priority. By comparison, in none of the research papers we have been citing does the argument against mesosomes proceed by adverting to the (negative) robustness of observed results—the microbiologists here don't argue that, because a number of (independent) research groups fail to reveal mesosomes, mesosomes therefore don't exist.

When we arrive at the 1980s, the experimental arguments against the reality of mesosomes become more thorough. Dubochet et al. (1983) argue for the artifactuality of mesosomes by noting that mesosomes are not observed when viewing unstained, unfixed, frozen-hydrated bacterial specimens (frozen-hydrated specimens are observed while frozen). The basis of their argument is their claim that 'unstained, amorphous, frozen-hydrated sections provide a faithful, high-resolution representation of living material' (1983, 387). What is distinctive about Dubochet et al.'s (1983) work is the detail with which they engage in justifying this claim:

> This [claim] is correct if we accept (i) that the bacteria have not been damaged during growth in the presence of glucose and during the short harvesting process, (ii) that we have demonstrated that the original hydration of the biological material is really preserved in the sections, and (iii) that either the sections are free of artifacts or the artifacts can be circumvented. (387)

They then proceed to justify (ii) and (iii). For instance, there will not be any chemical fixation artifacts since chemical fixatives were not used. Also, sectioning artifacts, they note, can be identified since 'such artifacts all have in common the property of being related to the cutting direction' (388). Leaving these justifications aside, however, it is clear that Dubochet et al.'s argument against the reality of mesosomes is based on their belief in the reliability of their chosen experimental regimen (i.e., examining unfixed, amorphous, bacterial sections through frozen-hydration). Roughly, their argument is as follows: Their frozen-hydration approach is reliable (a claim they make an effort to justify); mesosomes are not seen with this procedure; thus, mesosomes do not exist. This is again the sort of argumentative strategy used by Ebersold et al. (1981), Silva et al. (1976), and Fooke-Achterrath et al. (1974) to demonstrate the nonreality of mesosomes, and it is manifestly not a form of robustness reasoning.

As a final example, Hobot et al. (1985) argue against the existence of mesosomes on the basis of their freeze-substitution techniques by first citing similar negative results obtained by Ebersold et al. (1981) (who also used freeze-substitution) and by Dubochet et al. (1983) (who used frozen-hydration). They also mention the earlier negative results found by Nanninga (1971), Higgins and Daneo-Moore (1974), and Higgins et al. (1976) using freeze-fracturing but not with the goal of grounding a robustness justification for their no-mesosome result. Rather, Hobot et al.

(1985) consider freeze-substitution and frozen-hydration techniques to be a significant improvement over freeze-fracturing since freeze-fractures, they claim, can occur in such a way as to hide organelles and other structures. This effect with freeze-fractures had occurred in other cases (e.g., with tubular variants of phage T4; see Hobot et al. 1985, 970), and Higgins et al. (1976) had previously suggested the possibility that this might be happening with mesosomes. Freeze-substitution and frozen-hydration, conversely, avert this troublesome situation. Apparently, then, Hobot et al. (1985) are adopting the following argumentative strategy in justifying a no-mesosome conclusion: Freeze-substitution and frozen-hydration are the most reliable preparative measures one can use in examining bacterial ultrastructure; the testimony of these measures records the absence of mesosomes; thus, mesosomes do not exist. Once more, this is the sort of 'reliable process' reasoning we found in previous contra-mesosome experiments.

It is worthwhile to emphasize that Hobot et al. (1985) do not find any particular merit in generating results using freeze-substitution that agree with the results of less reliable techniques, such as freeze-fracturing. One would have thought that such agreement would be of value to them if robustness had been their chosen principle of experimental reasoning.

Let us then grant that the determination in the 1980s that the mesosome was an artifact was the result of experimental microbiologists using what I have termed reliable process reasoning. Again, it is a form of reasoning that, first, identifies a process that has the character of producing true reports with inputs of a certain kind (it is a 'reliable process') and, second, records that one has an input of this kind, leading to the conclusion that a produced report is truthful. To be sure, one might view such a characterization of how scientists reason to be somewhat mundane, even obvious. But that consideration should not stop us from appreciating how different such reasoning is from robustness reasoning. Robustness purports to establish the reliability of an observational process by noting the convergence of its results with the results of other, independent procedures and then infers the truth of this result. Reliable process reasoning, conversely, assumes the reliability of a process and then, on this basis, infers the truth of an observed result. Of course, with the latter form of reasoning, there is the key issue of how one should go about justifying the claim that the observational procedure under consideration is in fact reliable. Here, one of the main strategies microbiologists use in justifying the reliability of their experimental procedures is to identify empirical support for these procedures. The following are some examples of this approach.

We saw earlier that Silva et al. (1976) dismiss the reality of mesosomes on the grounds that mesosomes are found when bacteria are fixed using osmium. Osmium fixation, they claim, distorts the morphology of bacteria, a claim they defend on the basis of their observations that osmium leads to the lysis of protoplasts and the efflux of K+ ions from the cell. As they comment, 'The observed rates of K+ efflux indicate that OsO_4 is acting directly on the cytoplasmic membrane of the studied bacteria, causing a breakdown of its permeability' (102). Conversely, using UA as a first fixative has neither of these observable effects (103). Hobot et al. (1985) cite a similar problem: OsO_4 fixation, they submit, leads to artifactual nucleoid shapes since 'it has been found [empirically, by another researcher] that OsO_4 and aldehydes rapidly induce leakage of small cellular solutes, particularly of potassium', which, on their view, induces 'a rearrangement of the cellular content before the cytoplasm became cross-linked and gelled, and that this consequently [influences] the distribution of the areas containing the DNA plasm [i.e., nucleoid]' (967). Freeze-substitution, on the other hand, avoids this troublesome situation, which is a credit to its reliability as a preparative measure. Hence, freeze-substitution gives a more accurate picture of the shape and structure of nucleoids (e.g., nucleoids are more dispersed than they appear with osmium fixation) and, correlatively, it demonstrates the nonexistence of mesosomes which, on freeze-substitution, are absent. Thus, what we are seeing in the work of Silva et al. and Hobot et al. is that the pivotal assumptions pertaining to the reliability of experimental processes—osmium fixation leads to artifacts, whereas UA fixation and freeze-substitution do not—are justified on empirical grounds.

A similar observation can be made with Dubochet et al. (1983) who justify their use of 'unstained, amorphous, frozen-hydrated sections' on the grounds that 'the original hydration of the biological material is really preserved in the sections' (387). This fact they believe to be demonstrated 'by freeze-drying experiments which showed that the mass loss during freeze-drying was as expected for fully hydrated specimens'

(387). Again, regarding the possibility of freezing damage with frozen-hydrated specimens, they comment, '[Such specimens were] not divided into domains of pure ice or concentrated biological material' (388). They continue, 'This is not surprising since the crystalline order of water in amorphous samples, judged from the half-width of the diffraction rings, does not exceed 3 nm' (388). Again, the strategy of Dubochet et al. is to use empirical considerations wherever possible not only in justifying their theoretical pronouncements (here, that mesosomes are artifactual), but also in supporting the experimental procedures used in such justifications.

Nevertheless, it would be asking too much for experimenters to provide empirical justifications for their assumptions (about the reliability of their observational procedures as well as about related issues) in all cases. There is no doubt that scientists work in addition with assumptions of high philosophical abstractness for which empirical support would be meaningless, such as 'one should seek empirical support for one's views about the world' and 'the physical world is independent of one's mind'. One would also expect scientists to make use of various assumptions intrinsic to the field in which they are working, a sort of lore about their subject matter inculcated during their education and promulgated with like-minded colleagues. To give an example of this lore, consider the Ryter-Kellenberger (R–K) fixation method that was a standard part of experimental methodology in experimental microbiology starting in the late 1950s until the 1970s (Rasmussen 1993, 237). This method involves fixing a specimen in osmium tetroxide and then embedding it in a polyester resin, thus allowing it to be thinly sliced for electron microscopic study. The applicability and relevance of this method was assumed by many of the microbiological experimenters—but how was itself justified? In their pivotal paper, Ryter and Kellenberger (1958) argue that the R–K method reliably depicts the true state of specimens for a number of reasons (see Ryter and Kellenberger 1958, 603, and Kellenberger, Ryter and Séchaud 1958, 674). These include (a) this method is the only one that provides consistent, reproducible results for all the cells in a culture; (b) it exhibits a fine nucleoplasm for all bacterial species studied whereas prior methods presented nucleoplasms with varying structures; and (c) it displays the head of a T2 bacteriophage as 'perfectly polyhedral'. The first reason suggests that the reliability of a method is a matter of its consistency and reproducibility, or a matter of its 'pragmatic' reliability. Such a factor is perhaps a necessary condition for an experimental methodology since any methodology is unusable if its results are continuously variable. The second and third conditions set forth specific disciplinary assumptions about, first, the structure of a nucleoplasm and, second, about the characteristic shape of certain phage heads. Here, certain assumptions intrinsic to the state of the art in microbiological theory are playing a role in calibrating the reliability of an experimental method. Clearly, in more fully assessing the reliability of this method, microbiologists could cite the empirical grounding for these assumptions—but the unquestioned familiarity of these assumptions to many microbiologists would probably make this unnecessary. As a matter of expedience, experimenters will justify the reliability of their methods on the basis of certain assumptions that have, for their part, been 'black-boxed'—that is, made into disciplinary truisms. The R–K method was itself black-boxed for many years; it became, by rote, a tool for generating 'reliable' observations—to the detriment, we might add, of microbiological researchers who were mistakenly led to believe in the reality of mesosomes through the use of the R–K method.

These are some of the ways, then, by which microbiologists go about justifying the reliability of their observational procedures. Many of these ways are discipline specific, utilizing the shared background knowledge of similarly trained researchers. Often the support is directly empirical, showing how a procedure is consistent with other observed facts; never is the support a form of robustness reasoning, where it is simply claimed that a procedure generates the same result as an independent procedure. It is hard to believe that anyone would be convinced by such an argument, where a consensus could just as easily be due to similar preconceptions and biases as it could be due to both procedures being reliable.

We mentioned earlier on that Nicolas Rasmussen, analogously to how we have been arguing, doubts the role of robustness reasoning in the mesosome episode (once more, in contrast to Culp's position). However, he combines his doubt with a general skepticism about the ability of philosophers to adequately understand the rationality of scientific work. Such a skepticism would affect my approach as well, if it were successful, because the only difference between my account and Culp's is what we take to be the 'rationality' behind the rejection of mesosomes. So our final task in this chapter is to get a handle on Rasmussen's skeptical sociological perspective and to explain why it fails to derail a reliable process reasoning interpretation of the mesosome episode.

RASMUSSEN'S INDETERMINISM

Earlier we mentioned Rasmussen's critique of Culp's work on the grounds that, even if she is right that robustness reasoning is used by experimental scientists, such reasoning is nevertheless too abstract and works at 'too low a level of resolution', to be effective in deciding scientific controversies. Rasmussen (2001) expands his target to more than just robustness but to practically any philosophically inspired rule of rationality. He says about such rules (and here we can include reliable process reasoning as among them) that

> Although [they] can be found at work in the reasoning of scientists from a wide variety of fields, they are too vague and abstract to pick out unambiguously, and thus to justify, particular scientific practices because there are many ways of instantiating them. Furthermore, though it is not incorrect to say that these principles have long been important to scientists, talking about these principles as if they are understood and applied in a uniform and unchanging way obscures the heterogeneity and fluidity of methodology as practiced within any given field—a degree of flux which is readily observed by higher-resolution examination of science over time. (634)

To illustrate this 'methodological flux' of scientific work, Rasmussen cites Nanne Nanninga's experimental work on mesosomes from 1968 to 1973. The core methodological issue for Nanninga's work during this period, according to Rasmussen, is whether the ultrastructure of bacterial specimens is better preserved using freeze-fracturing with the use of a cryoprotectant (glycerol) or without. As Rasmussen rightly points out (2001,

640), Nanninga (1968) supports the use of glycerol. In Nanninga's words, when using glycerol,

> Two observations indicate that we have succeeded in obtaining fairly reliable preservation of the ultrastructure of our specimens. (a) The bacterial cells grown in the presence of glycerol and frozen at approximately –150°C resumed growth when inoculated into fresh heart-infusion broth; this is in accordance with the results obtained by [H.] Moor with frozen yeast cells. (b) No signs of plasmolysis were seen in thin sections of bacteria cultivated in broth supplemented with glycerol. (253)

Moreover, again rightly, Rasmussen indicates that Nanninga (1973) abandons the requirement of using a cryoprotectant, a change for which Nanninga provides 'grounds' (Rasmussen 2001, 640). However, Rasmussen ignores these grounds and instead remarks,

> Regardless of how the change may have been justified, intellectual method did shift and as a result so did the implications of one line of evidence. (640–641)

We may have an indication, then, for why Rasmussen sees only capricious flux in the change of scientific methodologies when we see him ignoring the higher resolution detail that would reveal methodological constancy. To understand this further detail, consider what Nanninga (1973) says about the use of glycerol as a cryoprotectant:

> Without a cryoprotective agent such as glycerol, the heat transfer between the object and the freeze-fracturing agent is rather inefficient resulting in comparatively slow freezing and the concomitant formation of large ice crystals. In consequence bacteria are frequently squeezed between the crystals. Structures observed are, for instance, triangles which bear little resemblance to the original rod-shaped. Ice crystals inside the bacterium are always smaller than on the outside. When the ice crystals have dimensions similar

> to cytoplasmic structures (ribosomes), the interpretation becomes especially hazardous. (154)

To this point Nanninga is reiterating the common worry with the formation of ice crystals and highlighting the associated benefits of glycerol. However, he continues,

> Fracture faces of membranes on the other hand are relatively unaffected by ice crystals. Increasing concentrations of glycerol promote the formation of smaller crystals and thus reduce mechanical damage. However, glycerol may have an osmotic effect. For instance, mitochondria in yeast cells appear rounded when frozen in the presence of glycerol. Increasing the freezing rate by high pressure and omitting glycerol preserves their elongated structure. (154–155)

The key point for us to emphasize in these passages is that Nanninga's judgment about glycerol—that it may lead after all to poor preservation of a specimen—is not arbitrary or capricious in the least: It is based on related observations concerning the appearance of mitochondria, in particular, that mitochondrial structure is found to be distorted when frozen with glycerol. The presumption here, of course, is that the true structure of mitochondria is already known and that a distortion of mitochondrial structure would have an analogue in bacterial ultrastructure. Given these facts, Nanninga is in essence suggesting that the use of glycerol, given its osmotic effects, leads to unreliable preservation and so should be avoided, whereas the omission of glycerol leads to more reliable preservation and a more accurate picture of ultrastructure.

So why does Rasmussen see so much 'indeterminacy' in Nanninga's work? Perhaps the issue being focused on by Rasmussen is this: On the one hand, certain considerations weigh in favor of the use of glycerol (smaller ice crystals are less disruptive), whereas on the other hand certain considerations weigh against the use of glycerol (the increased osmotic pressure caused by glycerol distorts mitochondrial structure). How does one, then, go about resolving such a methodological dispute when the alternative approaches seem so equally compelling? Where there is no established way to resolve such a dispute, as might have been the case given the state of electron microscopic technology at the time Nanninga was writing, should we agree with Rasmussen that the fine-detail resolution of this dispute is to a certain extent capricious, epistemologically speaking, and only resolvable by reference to '"interests" and the other favorite mechanisms of the "strong programme" sociologists of knowledge' (Rasmussen 2001, 642)?

Nanninga (1973), for his part, never squarely faces this indeterminism. Rather, he simply ignores the arguments he had given in 1968 in support of the use of glycerol, because the question of glycerol and its beneficial or harmful effects on specimens becomes a side issue for Nanninga. Nanninga's focus turns instead to osmium tetroxide and the question of whether it (and not glycerol) is disruptive of subcellular ultrastructure and leads to faulty preservations. Let's consider the background of the osmium tetroxide issue.

Nanninga (1968) and Remsen (1968) had revealed the presence of mesosomes in freeze-fractured bacteria prepared without the use of osmium tetroxide. (Nanninga had additionally used glycerol whereas Remsen did not.) But Nanninga (1968) had also discovered mesosomes using freeze-fracturing *with* osmium fixation. Further experiments by Nanninga changed the significance of these results. In particular, Nanninga (1971) noted the following:

> We...observed that in unfixed and freeze-fractured cells mesosomes, if present, never reached the size and complexity that they did in freeze-fractured fixed cells. In neither case were mesosomes observed in the periplasm. (222)

Indeed, the situation with young cells is even more dramatic:

> The observation that mesosomal membranes (in contrast to the plasma membrane) cannot be clearly demonstrated in young *B. subtilis* cells unless chemical fixation is applied before freeze-fracturing is rather unexpected. (222)

On this issue, Nanninga (1973) becomes even more definitive, extending the above observation to bacterial cells generally and not just to young cells:

> By comparing the occurrence of mesosomes in freeze-fractured cells and in cells which had been chemically fixed with osmium tetroxide *before* freeze-fracturing, [a] considerable difference was observed between the two cases. . . . Chemical fixation before freeze-fracturing gave results comparable to thin-sectioning whereas without chemical fixation few if any mesosomes were found. (163, his italics)

Nanninga (1973) never goes so far as to conclude that mesosomes are artifactual. But he is clearly on his way to this conclusion for the following reasons. The use of osmium tetroxide is an integral part of the R–K method, but it is not a required step for the successful deployment of freeze-fracturing—osmium tetroxide is only needed when preparing specimens for thin-sectioning (and here the R–K method is used). Thus, when subsequent experimentation using freeze-fracturing without osmium fixation failed to exhibit bacteria with mesosomes or at least exhibited fewer and smaller mesosomes, and when this was compared to the familiar situation in which osmium-fixed bacteria exhibited large, centralized mesosomes with both freeze-fracturing and thin-sectioning, the suspicion occurred to Nanninga that osmium tetroxide might be a disruptive factor, perhaps not so far as actually creating mesosomes but at least to playing a role in enlarging or displacing them. (To be exact, Nanninga's [1973] conclusion is that small, peripherally located mesosomes more accurately represent bacterial cell structure than large, centralized mesosomes.)

Accordingly, from the above we can draw a few conclusions. First, we can allow as basically correct Rasmussen's claim that there was a change in methodology exhibited in Nanninga's work from the years 1968 to 1973, a change regarding the status of glycerol cryoprotection. However, Rasmussen underestimates the fact that the change was not capricious but based on empirical observations regarding mitochondrial structure. In other words, Rasmussen's assertion that experimental work on mesosomes (such as Nanninga's) involves a flux of methodologies lacking a substantive epistemic rationale is not borne out in the experimental work he is examining. Although we can admit that there is some uncertainty on Nanninga's part as regards what methodology is best in investigating freeze-fractured bacterial cells, his overall reasoning is straightforward: Because osmium tetroxide is not needed as a preparative measure with freeze-fracturing, and because freeze-fracturing without the use of osmium tetroxide both with and without glycerol exhibits bacterial cells without large, centralized mesosomes—whereas the use of osmium tetroxide in freeze-fracturing (and in thin sectioning) produces large, centralized mesosomes—it is reasonable to conclude that osmium tetroxide has a tendency to generate artifacts. That is, what Nanninga is providing us with is an argument for the *un*reliability of a particular experimental methodology—here the unreliability of using osmium tetroxide as a fixative—and then deriving the conclusion that the testimony of this method (that there exist large, centralized mesosomes) is mistaken. He is, to put it another way, applying the converse of reliable process reasoning, further illustrating how reliable process reasoning can be applied in experimental work.

At this stage we should be clear that, without a doubt, social, political and other nonepistemic interests find a place in scientific, experimental work, as they do in all human activities. We should also be clear that the application of reliable process reasoning (as well as robustness reasoning) in a particular case is always somewhat variable—just as with Nanninga's work with glycerol as a cryoprotectant, reliable reasoning can work in opposite directions depending on what other assumptions one makes. What we are denying is that such methodological openness introduces an irrevocable element of fluidity and vagueness into the application of epistemic principles, as Rasmussen seems to think. Scientists like Nanninga when confronted with indeterminate results do not lapse into a consideration of what nonepistemic factors might resolve this indeterminancy. Instead, they look to acquire more empirical information as a way of increasing the reliability and precision of their work, just as Nanninga turned to examining the experimental results produced using osmium tetroxide. This process of increasing one's empirical scope has no natural endpoint—there will always be further elements of openness and vagueness to confront—but that is just the character of our epistemic predicament as finite creatures. For one to suggest, as Rasmussen does, and perhaps as some sociologists of knowledge do, that the limitedness of our empirical resources and the open-ended nature of our rational methods makes the incursion of nonepistemic factors a necessity is to ignore completely what scientists see themselves as doing. It may be that scientists, in thinking they are constantly on the lookout for new empirical facts to (objectively) resolve their theoretical disputes, are suffering from some sort of false consciousness, unaware of their dependence on the strong programme's 'favorite mechanisms'—but that is a profound psychological claim for which neither Rasmussen nor the strong programmers have any empirical evidence.

Chapter 3

The WIMP: The Value of Model Independence

In the previous chapter, we looked at a case in which it was argued in the philosophical literature (by Sylvia Culp) that experimenters use robustness reasoning to support the accuracy of their observational results. In turn, we illustrated how the same experimenters neither avowed the use of robustness in print nor used robustness reasoning to support their views (which was probably wise, since by applying robustness they would most likely be led to conclude that mesosomes are real, contrary to the eventual settled view of the microbiological community). In addition, we saw how scientists were more inclined to use a different sort of reasoning, which I termed 'reliable process' reasoning. From this perspective, one starts with the (often empirically justified) assertion that an observational procedure is reliable—that is, given inputs of a certain kind, the procedure typically produces truthful observational reports—and in applying this procedure to the appropriate inputs is led to the conclusion that a generated observational report is true.

In order to further support the claim that scientists are not prone to use robustness reasoning in the way some philosophers think they are, and to provide additional grounds for my claim that scientific observers are best interpreted as applying reliable process reasoning, I turn now to an entirely different area of scientific research, an episode in the recent history of astroparticle physics. The episode concerns the observational search for one of the main candidates for cosmological dark matter, the so-called WIMP (weakly interacting massive particle, theoretically understood as the neutralino, the lightest superpartner in the supersymmetric extension of the standard model of particle physics). The search for WIMPs has been, and currently is, an intense area of astrophysical, observational research, and below we review the work of four research groups intent on finding, if possible, a positive sign for the existence of WIMPs. One particular group, DAMA (DArk MAtter, based in Italy), claims to have found just such a sign, and we examine the grounds DAMA provides for its optimism. In understanding DAMA's reasoning, it is useful for us to distinguish, as DAMA does, between two broad kinds of observational research: model dependent and model independent. The former kind of research involves constructing observational procedures that are heavily invested in a variety of background ('model') assumptions. One might anticipate that this would be a negative feature of an observational procedure, to be so reliant on background assumptions, but a model-dependent approach has the virtue that, if these assumptions turn out to be justified, the resultant observations are highly informative and detailed. By contrast, a model-independent approach seeks to reduce the number of assumptions needed in generating observed results while still ensuring informative, observational results. Clearly if these results are informative enough, one will have succeeded in generating observations that can resolve a scientific issue in a way that minimizes the chance for error. In effect, in the case at hand, DAMA claims that using a model-independent observational procedure generates a positive indicator for the existence of WIMPs; additionally, it disregards the negative observational indicators regarding WIMPs that have been generated by the groups with which it is competing on the grounds that these approaches are (excessively) model dependent and so cannot be trusted.

Our discussion of this debate between DAMA and its competitors will serve two main goals. First, it illustrates once more how when we look at actual scientific practice we do not find robustness reasoning being applied. Indeed, we will find that such reasoning is overtly disavowed by two of the research groups we are considering. Second, it will become apparent how what I call reliable process reasoning is, despite its simplicity, a philosophically accurate way to understand the reasoning of these groups. In effect, a model-dependent observational procedure is unreliable because of its excessive dependence on a variety of assumptions (thus its results cannot be taken to be accurate), whereas a model-independent approach is preferable for just the opposite reason—its relatively thin dependence on background assumptions provides warrant for its reliability and the attendant accuracy of its observed results.

To get us started in thinking about this WIMP episode, let us begin by reviewing some of the scientific background to explain why astrophysicists think dark matter exists at all.

DARK MATTER AND WIMPS

Dark matter is matter that is undetectable by means of electromagnetic radiation but that acts gravitationally just like ordinary matter. This mysterious hypothetical substance is thought to make up about 25% of the total constitution of the universe (as compared to 5% for regular luminous matter, the matter that we see around us and that common opinion takes to make up the entirety of the universe; 70% is dark energy, yet another mysterious substance thought to be a form of 'repulsive gravity'). There are a number of reasons why scientists believe in the existence of dark matter. One reason is that the velocities of galaxies in large assemblages of galaxies (i.e., 'velocity dispersions in galaxy clusters') are much faster than would be anticipated given how much mass is observed to exist in a cluster, assuming the general principles of gravitational force common to both Newtonianism and general relativity, particularly the universal law of gravitation and second law of dynamics (for background, see Moffat 2008, 71–73, and Gates 2009, 22). These velocity dispersions are great enough to exceed the anticipated escape velocities of the galaxies, which means these clusters should be dissipating away and not, as is observed, maintaining their gravitational bond. In order, then, to restore the consistency of observation and gravitational theory, it is often assumed by astrophysicists that there is in galaxy clusters, in addition to the mass that we can see (i.e., luminous mass), extra mass that acts gravitationally just like ordinary matter but that is nonluminous in that it cannot be directly detected by means of light or any other form of electromagnetic radiation. This extra mass, or dark matter, explains why galaxy clusters stay together, and because of this explanatory ability it is inferred that dark matter exists.

A similar explanation is given for why the outer edges of spiral galaxies (galaxies that spin around their centre, such as with the Milky Way) rotate faster around the centre of the galaxy than would be predicted on the basis of similar gravitational assumptions. If the only mass in a galaxy is luminous mass, and assuming the same general principles of gravitational force, the velocities of stars at the outer periphery of a spiral galaxy should steadily decrease. But what we find are 'flat' rotation curves: The velocities of stars level off at the distant edge of a galaxy and only slowly decrease at much further distances. Once more, these anomalous observations can be explained by assuming the existence of dark matter (Moffat 2008, 73–74, and Gates 2009, 22–23). More theoretically speculative justifications for the existence of dark matter derive from the need to account for (a) the formation of light elements in the early universe (called Big Bang nucleosynthesis; see Gates 2009, 23–27, and Filippini 2005) and (b) the formation of large-scale structures such as galaxies and galactic clusters (see Gates 2009, 162, and Primack 1999, §1.1). Each of these occurrences, it is argued, is inexplicable without the postulation of dark matter. Taken as a whole these 'explanatory' justifications (or 'inferences to the best explanation') have convinced many astrophysicists of the existence of dark matter.

The justification for the existence of dark matter, we should note, is not without controversy, and in chapter 5 we look closely at a recent attempt to provide a more direct justification. For now, taking the reality of dark matter for granted, we examine research aimed at determining the constitution of dark matter, particularly research centered on one of the main theoretical candidates for dark matter, the WIMP (other candidates, not considered here, include axions and light bosons; see Bernabei et al. 2006, 1447).

DAMA'S MODEL-INDEPENDENT APPROACH

A number of astrophysical research groups are working toward the possible isolation and identification of WIMPs (or, more precisely, WIMP detector interaction events). One such research group, DAMA, claims to have succeeded at the task of tracking WIMPs, and its positive result has generated a lot of debate in the astrophysical community. The key feature of DAMA's approach to WIMP detection (or what DAMA [2008] prefers to call 'dark matter' detection—we retain the acronym WIMP for consistency) is that this approach (in DAMA's terms) is 'model-independent'. Roughly, DAMA's idea, which we examine below, is that to effectively identify WIMPs one needs to adopt a model-independent approach in the sense that the number of assumptions needed in an observational procedure is minimized.

The process of detecting WIMPs is a complex affair. In the WIMP detectors used by DAMA, detection occurs by the means of a process called 'pulse shape discrimination'. Here, incoming particles interact with the constituent nuclei of a target material, which is typically located deep in a mine (to filter out noise generated by other sorts of incident particles). The target material used by DAMA is the scintillating crystal NaI(T1) (thallium-activated or thallium-doped sodium iodide), which emits flashes of light when subatomic particles, such as WIMPs, muons, gamma rays, beta rays and ambient neutrons, interact with either the crystal's nuclei or electrons, causing them to recoil. The flashes produced by a recoiling NaI(T1) nucleus are distinguishable from the flashes produced by a recoiling Na(T1) electron in that they have different 'timing structures' (i.e., the intensity of the flash measured relative to the flash's duration exhibits a different curve dependent on whether we are considering the recoil of a neutron or an electron). Accordingly, because WIMPs cause nuclear recoils, whereas gamma and beta radiation cause electron recoils, one way to identify an incoming WIMP is to look for those flashes of light characteristic of nuclear recoils. Unfortunately, muons and ambient neutrons also cause nuclear recoils, so DAMA in its experimental set-up aspires to minimize the background contribution of muons and neutrons. For example, by performing its experiment deep in an underground mine, they significantly reduce the impact of incident muons. Still, as DAMA sees the situation, one can never be sure that one has correctly identified a detection event as a WIMP interaction—as opposed to a muon, neutron or some other type of interaction that can mimic a WIMP interaction—because of the enormous number of potential, systematic errors emanating from the surrounding environment that can affect the output of the detector. It would be ideal, of course, if we could separate out precisely the WIMP events, and research groups competing with DAMA, such as Expérience pour DEtecter Les Wimps En SIte Souterrain, (EDELWEISS, based in France) and Cold Dark Matter Search (CDMS, based in the United

States), attempt to do this. Such attempts DAMA describes as 'model-dependent': They attempt to isolate individual WIMP detection events and with that attempt burden the accuracy of its results with an excessive number of auxiliary assumptions. For this reason DAMA expresses skepticism about the potential for a model-dependent approach to generate reliable results, given the difficulty of such a case-by-case identification of WIMPs using the pulse shape discrimination method. They say that any approach that purports to distinguish individual WIMP-induced recoil events from other sorts of recoil events using timing structures

> even under the assumption of an 'ideal' electromagnetic background rejection, cannot account alone for a WIMP signature. In fact, e.g. the neutrons and the internal end-range α's [alpha particles] induce signals indistinguishable from WIMP induced recoils and cannot be estimated and subtracted in any reliable manner at the needed precision. (Bernabei et al. 1998, 196, fn 1)

One of the distinctive features of DAMA's own approach, which it calls 'annual modulation analysis', is that it bypasses the need to make case-by-case discriminations of WIMP detection events. This is possible in part because DAMA's model-independent approach itself 'acts…as a very efficient background rejection [device]' (Bernabei et al. 1998, 197; see also Bernabei et al. 1999, 451). We will see in a moment how its strategy achieves this result.

The particular model-independent approach to WIMP detection advocated by DAMA (i.e., annual modulation analysis) employs the following cosmological model. Our galaxy, DAMA asserts, is immersed in a WIMP halo that fills in the spaces between its luminous components (such as stars and planets). It is a halo whose existence is inferred partly on the basis of observations of the rotation curves of spiral galaxies: Our observations of these curves seem to imply that galaxies are immersed in an unseen, that is, dark, though gravitationally significant field of mass. Once we grant the existence of this halo, it follows that, as our solar system rotates around the galactic centre, we are subject to what might be termed a 'WIMP wind'. The velocity of this wind will vary with the time of year as the earth rotates around the sun, dependent on whether the earth, relative to the sun's (i.e., the solar system's) movement through the WIMP halo, is moving with the sun or away from the sun. With this cosmological perspective in mind, we gain a rough idea of how the incidence of WIMPs on the earth will vary over the course of a year—WIMPs (if they exist) will be observed to exhibit an 'annual modulation'. As a way of detecting this modulation, DAMA's strategy is to set up WIMP detectors that look for trends in the detected nuclear recoils without distinguishing between which recoils are caused by WIMPs and which are caused by such things as neutrons or muons. It follows that, in its recorded data, DAMA allows there to be a share of false positive events appearing in its detectors that wrongly indicate the presence of WIMPs. The idea is that if it turns out that these particle interactions exhibit an annual modulation as predicted by the above cosmological model, and if we further could not attribute this modulation to any other source, then we have an assurance that we are witnessing WIMP detector interactions without needing to specify directly which particular nuclear recoils are WIMP events and which are not.

According to DAMA, this is what it succeeds in doing. On the basis of its DAMA/NaI experiment, which ran for seven years up to 2002, and then on the basis of its improved DAMA/LIBRA experiment, which began in 2003 and (as of 2013) is currently running, DAMA has collected a large amount of experimental data that displays how the rate of nuclear recoils (or, more generally, single-hit events) varies throughout the year. There are yearly peaks and valleys corresponding to a theoretically expected June/December cycle, one that takes the shape of the theoretically predicted cosine curve. DAMA, in considering this result, does not see how this result could be due to any source other than cosmic WIMPs. In regards other causes of nuclear recoils, such as ambient neutrons or a form of electromagnetic background, DAMA states that 'it is not clear how [these factors] could vary with the same period and phase of a possible WIMP signal' (Bernabei et al. 1998, 198). For instance, despite taking extreme precautions to exclude radon gas from the detectors (Bernabei et al. 2003, 32, and Bernabei et al. 2008, 347–348), DAMA nevertheless looks for the presence of any annual modulation of the amount of radon that might, hypothetically, cause a modulation effect—and it finds none. Moreover, DAMA notes that even if radon did explain the modulation, this modulation would be found in recoil energy ranges beyond what is observed (i.e., not only in the 2 to 6 keV range but also at higher ranges), and this is also not found in the experimental data (Bernabei et al. 2003, 34, Bernabei et al. 2008, 340). Similarly, DAMA examines the possibility of hardware noise causing a modulation signal (Bernabei et al. 2003, 36–37, Bernabei et al. 2008, 348–349), and, leaving aside the lack of any indication that such noise has a yearly modulation cycle, there is not, it determines, enough noise to generate a signal. Assessments along these lines are also made with regard to temperature, calibration factors, thermal and fast neutrons, muon flux and so on, and in no case does it seem that any of these effects could reproduce the observed modulation effect.

We indicated above that DAMA describes its approach as model independent in that it seeks to reduce the number of assumptions that need to be made in exploring the existence of WIMPs. To a degree DAMA succeeds at this reduction because what it is seeking is something more general than individual WIMP detector events: It seeks only to find trends in the nuclear recoil data indicative of the existence of WIMPs and does not strive to pick out WIMP detection events individually. As a result, DAMA can dispense with a number of assumptions necessary to ensure that one is detecting a WIMP and not something, like a neutron, that mimics WIMPs. But the independence DAMA is claiming credit for goes further than this: Given the observed annual modulation in the nuclear-recoil events, DAMA rules out (as we saw) the possibility that this modulation could have been caused by such things as ambient neutrons, the electromagnetic background, radon gas, temperature, calibration factors and muon flux. Simply, it is difficult to see how these factors *could* produce a modulation effect. Thus, DAMA has a two-pronged strategy aimed at removing the influence of background conditions on its results: Not only does it take meticulous care at removing these background influences; it also generates a result that, even if there were background influences, would seem inexplicable on the basis of them. In this way DAMA's results are model independent: The results hold independently of the status of a number of background model assumptions.

Unfortunately for DAMA, its positive result for the existence of WIMPs is the target of dedicated critique by other groups working on WIMP detection. The United Kingdom Dark Matter group (UKDM, based in England), in addition to CDMS and EDELWEISS, all assert that, if DAMA is right about WIMPs, then they too should be seeing WIMPs in their own experimental data—and they don't. What is interesting for us is why DAMA finds these critiques unconvincing. Whereas DAMA does not seek individual WIMP identifications per se but seeks only trends in the data that are best explained by the existence of WIMPs (in this way its approach is model independent), these other groups do seek to make individual WIMP identifications and thus adopt what DAMA calls a model-dependent approach to detecting WIMPs. They are model dependent on DAMA's account because, relative to DAMA's own approach, their claims are correlatively more dependent on what assumptions they rely on (which follows from the fact that they are more ambitious in their goals). As such, DAMA criticizes the work of these groups as burdened by both a general uncertainty regarding the 'astrophysical, nuclear and particle physics assumptions' they need to derive their results, as well as by a lack of precision concerning various other 'needed theoretical and experimental parameters', such as 'the WIMP local velocity...and other halo parameters [such as]...form factors [and] quenching [factors]' (Bernabei et al. 2003, 8). Indeed, these other approaches are so model dependent that their experimental conclusions, DAMA claims, 'should be considered only strictly correlated with the "cooking list" of the used experimental/theoretical assumptions and parameters' and thus have 'no general meaning, no potentiality of discovery and—by [their] nature—can give only "negative" results' (9). As DAMA summarizes its concern, such model-dependent experiments

> exploit a huge data selection...typically [involving] extremely poor exposures with respect to generally long data taking and, in some cases, to several used detectors. Their counting rate is very high and few/zero events are claimed after applying several strong and hardly safe rejection procedures.... These rejection procedures are also poorly described and, often, not completely quantified. Moreover, most efficiencies and physical quantities entering in the interpretation of the claimed selected events have never been discussed in the needed [detail]. (21)

To help us see the point of DAMA's critique, let us examine some of the work of these other groups.

MODEL-DEPENDENT APPROACHES TO DETECTING WIMPS

To begin with, it's worthwhile to point out that all the participants to this experimental controversy use, roughly, the same methodology in tracking the existence of WIMPs. They each set up a shielded detector located deep in the ground (sometimes at the bottom of mine shafts), a detector that has the capability of distinguishing between nuclear recoils (which are characteristically caused by WIMPs, neutrons and muons) and electron recoils (characteristically caused by gamma and beta radiation) as they occur inside the detector. Of the experiments we are looking at, two sorts of detection strategies are used. First UKDM, much like DAMA, uses a 'scintillation' approach in which a detector composed of NaI (sodium iodide) emits flashes of light (scintillations) when bombarded with subatomic particles. Once again, depending on which kind of subatomic particle we are dealing with, each such WIMP detector interaction has a distinct form of scintillation that is picked up by photomultiplier tubes (PMTs) viewing the detector. On its pulse shape discrimination approach, UKDM focuses on the 'time constant' of a scintillation pulse (in essence, the time when the pulse is half-completed); nuclear recoils have characteristically shorter time constants, whereas electron recoils have longer ones. Comparatively, CDMS and EDELWEISS use a 'heat and ionization' approach based on the principle that nuclear recoils are less ionizing than electron recoils. As such, the 'ionization yield'—the ratio of ionization energy (the amount of charge generated by a recoil) to recoil energy (the total energy produced by a recoil)—is smaller for nuclear recoils (which again could be caused by prospective WIMPs) than it is for electron recoils.

From 2000 to 2003, UKDM operated a sodium iodide, scintillation detector in the Boulby mine in the UK in an experimental trial called NAIAD (NaI—sodium iodide—Advanced Detector; see Alner et al. 2005, 18). Using pulse shape discrimination, UKDM examined the time constant distributions for scintillation pulses for two cases: case (a) examining the distribution that results from exclusively gamma radiation (gamma rays cause electron recoils) and case (b) which exhibits results for both electron and nuclear recoils (where such nuclear recoils could be caused by incident muons, neutrons or WIMPs). As time constant values for nuclear recoils are generally smaller than those for electron recoils, with case (b) we'd anticipate seeing events with smaller time constant values than we normally see in case (a). In fact this is exactly what we do see, indicating the occurrence of nuclear recoils and thus possibly of WIMPs. However, UKDM ascribes these short time constant events to PMT noise—in effect, the PMTs that pick up scintillation light from crystals generate their own information that mimics nuclear recoils. As a result, UKDM performs the relevant 'cuts', excluding the photomultiplier 'background', and arrives at a corrected curve that looks practically identical to the pure gamma ray (calibration) curve. From here UKDM concludes, 'No contribution from WIMP-nucleus interactions were [sic.] observed in these data' (Alner et al. 2005, 22). That is, any events it might have identified as WIMP events were written off as photomultuplier background noise.

DAMA acknowledges that, to find a WIMP signal in the way UKDM does, one needs to account for possible sources of error that might misleadingly mimic this signal. Yet DAMA's concern is that groups like UKDM have set themselves too difficult a task in isolating individual WIMP interaction events. Because such approaches are model dependent, as DAMA calls them, there are a large number of factors that need to be considered to retrieve informative results. As a result, in accounting for these factors, these groups must cut back, sometimes to extraordinary lengths, on potentially perceived pronuclear recoil/pro-WIMP data events. We've noted, for instance, the cuts UKDM needs to make to account for PMT noise. Let us now look at the work of another group that takes a model-dependent approach.

CDMS operates heat and ionization Ge (germanium) and Si (silicon) detectors deep in a mine in Minnesota, and it provides an extensive and impressive tabulation of the various data cuts that need to be made to properly isolate a WIMP signal. In this regard, starting from 968,680 possible WIMP detection events, CDMS proceeds with data cut after data cut and eventually ends up with one event, which is itself eventually dismissed as having an occurrence 'consistent with our expected (surface) electron-recoil misidentification' (Akerib et al. 2005, 052009-35). CDMS makes these cuts on the grounds that, on its estimation, the detector at hand has the unfortunate feature of producing data that can mimic WIMP events. For instance, one of the cuts involves the fact that only nuclear recoil events involving scattering in a single detector are used (in CDMS's experimental set-up, a number of detectors are used simultaneously); WIMPs do not multiply scatter, and so only single scatter events need to be counted. Again, CDMS uses a cut called 'muon veto', which refers to the fact that nuclear recoils can occur as a result of incoming muons, and so the detector is shielded by a muon veto made of plastic that is set off by the presence of an incoming muon. Hence, when the veto indicates the presence of a muon coincident with the occurrence of a nuclear recoil in the detector, the nuclear recoil is discarded as a possible candidate WIMP event. All the numerous cuts CDMS makes are of a similar nature—in essence, it specifies possible sources of 'false positive' events and thus forms a basis on which to discard data. Eventually all the possible WIMP detection events are discarded on the basis of these cuts, from which CDMS proceeds to conclude that no WIMP interaction events are seen (see Akerib et al. 2005, 052009-34).

At this stage one might commend CDMS for its vigilance in discarding possible erroneous WIMP detection events. CDMS might here be thought to be expressing only warranted prudence, a careful skeptical attitude that rejects dubious (or potentially dubious) 'hits' in order to achieve a high degree of probability when a positive event is claimed. Given the importance a positive detection would have, doesn't this sort of prudence seem appropriate rather than problematic? However, DAMA takes a very different view of the matter. In reflecting on such model-dependent approaches, DAMA notes 'the existence of known concurrent processes…whose contribution cannot be estimated and subtracted in any reliable manner at the needed level of precision' (Bernabei et al. 2003, 10). Some of these 'concurrent processes' were listed above, that is, muon events and multiple scatterings. DAMA highlights as well what are known as 'surface electron' events. It had been noted in both pulse shape discrimination experiments (e.g., by UKDM in Ahmed et al. 2003) and in heat and ionization experiments (e.g., by EDELWEISS in Benoit et al. 2001 and by CDMS in Abusaidi et al. 2000) that there is a set of events occurring near the surface of a detector in both sets of experiments that is able to effectively mimic nuclear recoils (and thus potential WIMP events). As a result, to meet the challenge of such surface electron events, various measures are put in place to exclude such events: UKDM uses unencapsulated crystals instead of encapsulated ones (Ahmed et al. 2003, 692), CDMS goes so far as to discard a detector that exhibits an excess of such events (Abusaidi et al. 2000, 5700), and EDELWEISS restricts its data gathering to a 'fiducial volume of the detector' (roughly, the centre part of the detector as opposed to its outer edge—see Benoit et al. 2001, 18). DAMA's concern, as expressed in the above quote, is that, whichever method one uses, one possibly discards genuine nuclear recoils and thus possibly discards WIMP detection events as well. All that might be just fine if we knew exactly what was occurring in these experiments—but DAMA doubts that we do and thus rebukes the excessive caution expressed by the other groups.

Similar to CDMS, EDELWEISS utilizes heat and ionization experiments exploiting the phenomenon that nuclear recoils are less ionizing than electron recoils (see Di Stefano et al. 2001, 330, Abrams et al. 2002, 122003-2, and Akerib et al. 2004, 1, for discussion of this point). For nuclear recoils, the ratio of ionization energy (i.e., the amount of charge generated) to the recoil energy (i.e., the total energy produced by a recoil) is less than the corresponding ratio for electron recoils (the ionization yield). In identifying WIMP interaction events by this means, there are two issues to consider: (a) how to distinguish WIMP interaction events (i.e., nuclear recoils) from electron recoils, and (b) how to distinguish the nuclear recoils caused by WIMP interaction events from the nuclear recoils caused by other sorts of interactions (i.e., involving mainly incident muons and ambient neutrons). Step (a) is fairly straightforward: The ionization yields for electron and nuclear recoils are clearly distinct. But step (b) is more contentious, and, once more, many procedures are deployed to isolate WIMP events from other sorts of nuclear recoils, such as installing thick paraffin shielding to absorb incident neutrons, using circulated nitrogen to reduce radon amounts, retrieving data from only the 'fiducial' volume and so on (Benoit et al. 2001, 16). Taking into consideration as well the need to account for its detector's efficiency (on efficiency, see Sanglard et al. 2005, 122002-6), EDELWEISS then concludes that there are no WIMPs observed at a 90% confidence level. This result, EDELWEISS infers, refutes DAMA's claimed WIMP modulation signature that supports the existence of a WIMP modulation signature.

What is interesting to note is that, in later, 2003 experimental work (described in Sanglard et al. 2005), EDELWEISS improves its apparatus in ways that increases its receptivity to nuclear recoil events (e.g., by reducing background noise) and that increases its efficiency at lower energies (Sanglard et al. 2005, 122022-6). The result is that EDELWEISS arrives at a total of 59 WIMP candidate events (i.e., nuclear recoils). This is a much more substantive result, and one would think that in this preponderance of data one might find some true WIMP candidate events. But that is not how EDELEWEISS interprets the data: Rather, it cites various problematic sources of contaminating background information, particularly 'bad charge collection of electron recoils near the surface of the detector [i.e., surface events]' and residual neutron flux in the detector's ambient environment' (122002-13), and concludes that

> in the absence of more detailed studies, it is not possible to conclude quantitatively [about the extent of these contaminating sources] and therefore no background subtraction is performed for the estimate of the limits on the WIMP collision rate in the detectors. (122002-14)

One would think that such a pronouncement would put an end, temporarily, to the investigation, pending a more adequate accounting of these sources of error. But EDELWEISS is unperturbed: It recommends using the 'optimum interval method' suggested by Yellin (2002) that is 'well-adapted to [its] case, where no reliable models are available to describe potential background sources and no subtraction is possible' (Sanglard et al. 2005, 122002-14). Adopting this method leads EDELWEISS to a result largely in line with its 2002 assessment: Whereas in 2002 it finds no nuclear recoils (above the 20 keV threshold), it now finds three nuclear recoil events, which is consistent with the previous result given the proportionately longer exposure time on which the latter data is based. On this basis, EDELWEISS draws a conclusion that, again, refutes DAMA's modulation signature.

At this stage, one might find oneself sympathizing with DAMA regarding its bewilderment about the argumentative strategies adopted by 'anti-WIMP detection' experimental groups such as EDELWEISS. The

Yellin approach is highly idiosyncratic and is not used anywhere else in the WIMP detection literature; moreover, it is obviously no substitute for an experimental approach that, instead of conceding the absence of 'reliable models...available to describe potential background sources' (Sanglard et al. 2005, 122002-14), takes steps to account for or (even better) remove interfering background information. In this regard EDELWEISS in the concluding section of Sanglard et al. 2005 (after utilizing the Yellin method) describes its plan to improve its detectors, increasing their size and numbers. Moreover, it notes that it has plans to drastically reduce the problematic neutron flux by '[installing] a 50 cm polyethylene shielding offering a more uniform coverage over all solid angles' and to also utilize 'a scintillating muon veto surrounding the experiment [that] should tag neutrons created by muon interactions in the shielding' (Sanglard et al. 2005, 122002-14). From the perspective DAMA adopts, these sorts of measures need to be put in place before EDELWEISS can draw any conclusions denying the existence of WIMP detection events, particularly where such a denial is based on an admission that there is background information that cannot be reliably accounted for. EDELWEISS candidly admits that there is both a lack of clarity about which events are nuclear recoil events and significant uncertainty in picking out from a set of nuclear recoil events those events resulting from WIMP detector interactions. As DAMA expresses the problem, the WIMP identification strategies of EDELWEISS, CDMS and UKDM are model dependent because of their reliance on a multitude of difficult-to-ascertain model assumptions, and for this reason their work is unreliable. Better, DAMA thinks, to adopt a model-independent approach, and, as we have seen, this approach leads us to isolate a WIMP annual modulation signature.

AN HISTORICAL ARGUMENT AGAINST ROBUSTNESS

In the historical case we are examining, we have seen how various research groups have argued against DAMA's positive WIMP identification by attempting to identify individual WIMP interaction events. What is interesting for us is how these groups completely ignore the strategy of deploying robustness reasoning, despite its seeming usefulness in assuring an experimental conclusion. Particularly, all three anti-DAMA groups (UKDM, CDMS and EDELWEISS) retrieve the same negative result—none of them find a WIMP signal. Moreover, all three groups use experimental approaches that differ in various ways. For example, UKDM uses a scintillation detector, whereas CDMS and EDELWEISS use heat and ionization detectors; all the experiments occur in different countries and in different mines; and they all use different target masses with different exposure times. Thus, one might expect such groups in their published articles to argue in 'robust' fashion and to argue that because they all arrived at the same negative result, despite differences in their experimental methodologies, this negative result must therefore be correct. But this is not the case: None argue for their negative results by affirming its agreement with the other negative results retrieved by the other approaches. Instead, we find each of them arguing that its particular results are reliable insofar as it takes into consideration various sources of error—for instance, a group may argue that its results are more reliable because a muon veto was installed to account for the muon flux, or because a lead shield is present to protect the detector from the neutron background, or because the influence of photomultiplier noise is adequately accounted for and so on. In fact, these contra-DAMA groups sometimes squabble among themselves on points of experimental error. For instance, EDELWEISS found the work of CDMS in the shallow Stanford site to be problematic because it didn't effectively shield the detector from cosmic muons (Benoit et al. 2002, 44). Here, one might suggest that the application of robustness reasoning is inapplicable since we are looking at a convergence of negative results, but there is no such restriction on the robustness principle as it is usually expressed in the literature. In fact, we saw Culp use negative robustness in her interpretation of the mesosome episode.

Thus, for all its vaunted value in the philosophical canon, robustness does not appear to be much of a factor in the WIMP detection case we are currently considering. WIMP experimental researchers appear to eschew the sort of reasoning that runs thus: We generated this (negative) experimental result, as did these other experimental groups using different experimental approaches; thus, our experimental result is more likely to be accurate. Rather, such experimenters are much more focused on improving the reliability of their own experimental regimes (by removing background influences, ensuring the proper functioning of their apparatus and so on) even to the point of confuting the experimental value of other experimental approaches that arrived at the same result. One might potentially explain this resistance to robustness reasoning on the grounds that robustness is uninformative where the suggested, alternate forms of experimental inquiry are not reliable—what benefit is there to multiplying unreliable experimental routes? However, explaining the resistance to robustness reasoning in the WIMP case is not so easy. Although EDELWEISS questioned the reliability of CDMS's work, CDMS had no such similar complaint regarding EDELWEISS, and UKDM neither objected to, nor was criticized by, the other model-dependent approaches. In this historical case robust forms of reasoning were ignored, even when the reliability of alternate experimental routes was not subject to doubt.

Another historical consideration deriving from this episode that weighs against robustness involves a reflection on the methodological comments WIMP detection researchers make when they compare their methodologies to the methodologies adopted by other researchers. Specifically, we find them openly disavowing the requirement of robustness. Consider the following two sets of comments, the first from UKDM, which argues against DAMA's annual modulation result:

> Although several existing experiments have a potential to probe the whole region of WIMP parameters allowed by the DAMA signal (see, for example, [experiments performed by CDMS and EDELWEISS] ...), they use other techniques and other target materials. This leaves room for speculation about possible uncertainties in the comparison of results. These uncertainties are related to systematic effects and nuclear physics calculations. Running an experiment, NAIAD, with the same target (NaI) and detection technique but different analysis would help in the understanding of possible systematic effects. Such an experiment will also be complementary to more sensitive detectors in studying regions of WIMP parameter space favoured by the DAMA positive signal. (Ahmed et al. 2003, 692)

Here, we find UKDM explicitly disavowing any particular benefit in retrieving the same results as other groups, for these other groups use 'other techniques and other target materials' that, for UKDM, only increases the 'uncertainty' in the experimental data. Of course, UKDM knows that these other techniques yielded the same negative results as it does. But such robustness considerations don't seem to be a factor for UKDM. Better, the group thinks, to use 'the same target (NaI) and detection technique' with a 'different analysis', an approach it considers more informative.

DAMA, too, makes similar anti-robustness comments:

> Let us remark that the safest strategy is to compare results on exclusion plot and modulation obtained within the same experiment. In particular, the comparison of exclusion plots obtained by different experiments requires a consistent use of astrophysical (local density, velocities) and nuclear physics (matrix elements, spin factors, form factors) parameters. Also the instrumental effects (energy threshold, noise rejection capability, detector resolutions and quenching factors) have to be always adequately introduced. Moreover, for different target detectors further uncertainties could also arise because of the needed rescaling from the cross section of the different target-nuclei to σ_p (the WIMP-proton elastic cross-section) and because of possible different unknown or underestimated systematic errors. (Bernabei et al. 1998, 196)

Here DAMA is making the same methodological point made by UKDM: Better, it thinks, to focus on one experimental route (and to presumably work on improving its reliability, such as we find DAMA and the other experimental approaches doing, introducing 'improved' versions of their experiments year after year) than to start making comparisons with other experimental approaches that require the 'consistent use of astrophysical... and nuclear physics... parameters', that introduce 'instrumental effects' and that raise the possibility of 'further uncertainties' and 'different unknown or underestimated systematic errors'.

Now if DAMA were a proponent of robustness, it would have to compare its results with those of UKDM, CDMS and EDELWEISS, and this would certainly be problematic for its own perspective given that these other results conflict with its own. But DAMA's reasoning in the above quote indicates why it finds this approach problematic, and it is clearly reasoning that is not purely self-serving: These other approaches, because of their differences, simply raise more experimental questions than it is worth having. As we saw, UKDM argues in a similar fashion: Multiplying observational approaches 'leaves room for speculation about possible uncertainties in the comparison of results' (Ahmed et al. 2003, 692).

RELIABLE PROCESS REASONING

It appears, then, that in the episode we are considering researchers did not find much use for, and were even prone to be critical of, robustness reasoning. The way to understand this resistance, I submit, is to look at their methodological commitments in terms of an allegiance to reliable process reasoning.

Consider the resistance expressed by UKDM and DAMA to examining alternate observational procedures: Their worry was that doing this simply increased the uncertainty of the results. We can understand this if we view these scientists as seeking reliable observational procedures, procedures that are more likely to generate truthful results. The greater number of assumptions that need to be made for a procedure to work, the more prone this procedure is to error. Thus, for example, if we are supporting an observed result with two observational procedures that carry independent sets of background assumptions, and we plan to argue robustly and accurately, we need to assume the truth of both sets of assumptions. Particularly where our research is novel and more speculative, as it is with the search for (hypothetical) WIMPs, robustness only serves to complicate our investigations for we essentially multiply the assumptions we need to get right. Note that here we are thinking epistemically, as opposed to pragmatically, as Wimsatt does (see chapter 1). Robustness is valuable if we want to strategically support an observed result and are not concerned with the accuracy of the independent assumptions we need to make. Pragmatically, it's useful to have redundant support for a result. Apparently, then, UKDM and DAMA are not thinking in these terms when they overtly disavow robustness reasoning in the above quotes—they must be viewing their respective research tasks in epistemic, truth-tending terms.

But could there be other reasons why these research groups neglect robustness reasoning and even occasionally dismiss the value of a potential convergence of observed results using their relatively different observational procedures? Here one might cast sociological (or other external) explanations for why research groups prefer not to allude to the convergent results of other groups. For example, these groups may be in competition and may want to establish their priority in generating a result; alternatively, the members of a particular group may not be suitably positioned to comment authoritatively on the scientific value of another group's research and so are hesitant to make use of the results of this other group; indeed, the motivations of the researchers need not even be pure—one group may simply not want to be 'associated' with another group, despite their convergent data. Given these sorts of reasons, it need not follow that the resistance of a research group to robustly argue in the context of convergent data from another research group is a sign that this group does not recognize the epistemic value of robustness reasoning—perhaps it does, but these other external factors override the recognition of this epistemic virtue.

There is no doubt that such factors could be influencing the judgments of the research groups in this case and that, in such an event, using the above quotes to justify the claim that astrophysical researchers fail to see the point of robustness reasoning would be somewhat premature, pending a more thorough social scientific inquiry into the dynamics of the interactions between these groups. Still, there is reason to require here that any such external investigation be motivated empirically before it is taken seriously. This is because the internal, epistemic reading I have suggested—that these groups fail to see the epistemic value in multiplying observational 'angles'—falls very naturally out the details we have presented so far about the case. For instance, a presumed competition between the model-dependent groups (that hinders them from alluding to each other's work) is unlikely, given that what is retrieved is essentially a non-result—the nonidentification of a WIMP. There's no special priority in generating that sort of negative result since the vast majority of results are negative—DAMA's positive result is quite unique. Basically what the astrophysical community is doing is working on improving their detection devices, making them more sensitive, with improvements occurring all the time. As such, any presumed 'priority' would be extremely short lived. Moreover, UKDM's and DAMA's stated reasons for being hesitant to use the research results of other groups boils down essentially to the matter of the 'uncertainty' inhering in the procedures used by the other groups, not to the problem of being 'ignorant' of what these other groups are doing or lacking the wherewithal to properly understand these procedures. For robustness to apply, one need not have a comprehensive knowledge of how an alternative procedure works. One need only be assured that the other approach is indeed different and at least minimally reliable—and it is the dubious reliability of other approaches, from their perspective, that informs the judgments of UKDM and DAMA. Finally, and truly, if a research group dismisses the convergent results of other groups not because this group fails to recognize the value of robustness reasoning but simply because it harbours an irrational bias toward these other groups based on pure prejudice, then I think we should judge the research of the former group in a fairly bad way. This is not to deny that such attitudes might occur in science—only that such occurrences would amount to a sad abandonment of epistemic ideals and therefore would not be our concern.

Overall then, my inclination is to take the quotes from UKDM and DAMA at face value, as expressing an apprehension regarding the reliability of the work performed by other groups or at least expressing a resistance to engage in a detailed inquiry that thoroughly assesses the reliability of this work. This is surely not an unreasonable attitude for UKDM and DAMA to take, given that they are preoccupied with their own highly complicated research programs and given also that (as DAMA suggests) the observational procedures of these other groups are themselves model dependent and so dependent on a very large body of assumptions. Because model-dependent approaches are so heavily burdened by assumptions, it follows that applying robustness reasoning and showing that the same result holds while varying a few parameters does little to lessen the negative impact of model dependence. For example, consider that UKDM, CDMS and EDELWEISS all retrieved results initially supportive of the existence of WIMPs and did so by different routes (e.g., in different mines, in different countries, sometimes using different detector materials and so on). This robust convergence, nevertheless, is ineffective at countering the readiness with which these (model-dependent) groups discount presumably positive results. Each group potentially witnessed WIMPs and so had the basis on which to ground a robustness argument on behalf of WIMPs, but none of them argued along these lines because each group identified key sources of error in their experimental methodologies: UKDM with PMT noise; CDMS with muon events, surface electron events, multiple scatterings and so on; and EDELWEISS with the bad charge collection of electron recoils near the surface of the detector, residual neutron flux and other problems. Such errors persist and are decisive for these groups, irrespective of any robustness argument that might be formed using their convergent positive indicators of WIMP detector interactions. Because of the numerous and controversial assumptions at work in these model-dependent experiments, DAMA describes these research groups as working with 'cooking lists' of 'used experimental/theoretical assumptions and parameters' (Bernabei et al. 2003, 9). The groups can, in effect, cook up negative results without much effort. So in understanding what, in particular, the anti-DAMA groups are up to, reliable process reasoning (and not robustness) is perfectly apt: These groups recognize that their observational methodologies contain flaws and so are unreliable, which means that any positive result on behalf of WIMPs can be ignored. More than anything else, these groups are intent on improving the reliability of their experimental regimens in an effort to successfully identify individual WIMP interaction events. That is, they're not necessarily concluding that WIMPs don't exist—only that they haven't yet located an adequate experimental proof. It is this experimental uncertainty that grounds their misgivings over the value of DAMA's model-independent proof.

Similarly, DAMA's suggestion to reduce the number of assumptions needed in generating reliable experimental data—that is, to adopt what DAMA calls a model-independent approach—makes a lot of sense if we think in terms of reliable process reasoning. With a reduction in the number of the assumptions needed in using an observational procedure we proportionately reduce the risk of error. This is, in fact, particularly good advice in an area of inquiry where the subject matter is extraordinarily complex and our understanding is at a primitive stage, such as with dark matter research. On DAMA's view, it is better to pursue an inquiry that is less ambitious and that avoids overly precise discriminations of the phenomena being investigated than to engage in a more ambitious project that has little chance of generating a positive result due to its dependence on a controversial set of assumptions. In other words, DAMA places a premium on having a reliable experimental process, one that reduces the risk of error. With this reliable process in place, and due to its extensive empirical work in demonstrating the reliability of this process when it does in fact issue a positive WIMP interaction report, DAMA feels comfortable in asserting the truthfulness of this report, despite what appear to be robust negative results emanating from its competitors.

But couldn't we still find a place for robustness reasoning in DAMA's methodology, despite the fact that it is using model-independent procedures? For instance, we saw how DAMA in arriving at its annual modulation signature took precautions to exclude the contamination of its detectors by radon gas. These precautions aside, DAMA also argued that its annual modulation result holds even if the precautions were ultimately unsuccessful, since such an annual modulation cannot be explained by the presence of radon gas (i.e., if radon gas did produce a modulation, such a modulation would not match the modulation that was actually observed). Now suppose DAMA constructed a robustness argument along the following lines. It identifies two observational procedures: observational procedure A, in which radon gas is excluded and subsequently an annual modulation is witnessed, and observational procedure B, in which radon gas is not excluded but with the same annual modulation being witnessed. Let us assume that A and B are independent observational procedures (which is admittedly a questionable assumption given how much the two procedures have in common). Would there be a compelling robustness argument here, leading to the conclusion that the annual modulation result is not an artifact of the presence of radon? I think it is clear that we should not find this argument compelling: There is no direct merit in intentionally utilizing an observational procedure that involves a clear, possible source of error (such as when we allow the influence of radon). In this case, procedure A would have obvious authority, and procedure B would acquire its authority by having retrieved the same result as A. In effect, B is being calibrated by A (a strategy we will see utilized by Jean Perrin in the next chapter). When it comes to the situation with radon, the responsible action to take is to simply remove the possible source of error, which is what DAMA did in its observational procedure. More abstractly, if we know that an observed process has an established source of error, there is no added value in using data from this process to understand a phenomenon, if we have at hand a process that is the same except that it physically removes this source of error. This latter process is even better than a process in which the source of error isn't physically removed but is 'corrected for' in the final results. Including a physical error in an observational procedure and then conceptually correcting for it is less reliable than simply removing the physical error to begin with—it adds two steps, allowing an error and then correcting for it, to get back to where we started (which simply physically removes the source of error). Thus there is no worthwhile robustness argument here based on a convergence of observational procedures A and B—A is simply more reliable and should be the sole basis for one's observational conclusions.

Chapter 4

Perrin's Atoms and Molecules

For many philosophers (such as Cartwright 1983, Salmon 1984, Kosso 1989 and Stegenga 2009), the classic expression of robustness reasoning in the sciences is Jean Perrin's early 20th-century work in support of the reality of atoms and molecules. Perrin's arguments for the 'discontinuous structure of matter' (as he calls it in his 1926 Nobel Prize lecture) are set forth in two (translated) books, *Brownian Movement and Molecular Reality* (Perrin 1910) and *Atoms* (1916, 4th edition, and 1923, 11th edition), as well as in his Nobel Prize lecture (1926). Notably, Perrin portrays himself in these books and in his Nobel lecture as reasoning 'robustly' (though he doesn't use this more modern term): A key part of his proof of the reality of atoms and molecules is establishing an accurate value for Avogadro's number, and Perrin is explicit that his success at this task is due to the convergence of a variety of different physical processes that all lead to approximately the same number. Perrin's work thus poses a clear challenge to the critic of robustness: As one of the acknowledged paradigms of scientific reasoning, it apparently makes heavy use of robustness, and the author of this reasoning is overtly conscious of this fact.

My plan in this chapter is to determine whether it is really the case that Perrin uses robustness reasoning—his avowals that he is notwithstanding. This will involve us in a scrupulous reading of Perrin's writings, a reading that reveals Perrin's reasoning to be somewhat different from robustness. In particular, of the various experimental approaches that, according to Perrin, lead us to a value for Avogadro's number, one approach in particular—his vertical distribution experiments using emulsions—possesses for Perrin a degree of epistemic authority unmatched by the other approaches and so is a standard by which to 'calibrate' (or test) these other approaches. Thus, although it is true that Avogadro's number can be separately derived within an approximation by means of differing physical processes, the grounds for the accuracy of this number is not this independent convergence but rather the fact that this number is generated by the preferred approach. By then generating numbers in sync with this preference, the other approaches along with their theoretical underpinnings are thereby 'verified' (to adopt Perrin's term). A key virtue of reading Perrin in this way is that it provides an interesting explanation for why he believes his experimental work justifies a realism about atoms and molecules. This explanation will become useful at the end of the chapter in rebutting arguments advanced by Bas van Fraassen and Peter Achinstein, who claim that Perrin's realism is unfounded.

PERRIN'S TABLE

At the end of Perrin (1910), Perrin (1916) and Perrin (1923), a table is provided that summarizes the various physical procedures Perrin has either himself deployed or cited in deriving values for Avogadro's number (symbolized by *N*). To guide us in our examination, we focus on the table as presented in the English translation of the 4th edition of *Les Atomes* (1916). Perrin comments,

> In concluding this study, a review of various phenomena that have yielded values for the molecular magnitude [i.e., Avogradro's number, designated *N*] enables us to draw up the following table:

Phenomena observed	$N/10^{22}$
Viscosity of gases (van der Waal's equation)	62
Brownian movement—Distribution of grains	68.3
– Displacements	68.8
– Rotations	65
– Diffusion	69

Phenomena observed	$N/10^{22}$
Irregular molecular distribution—Critical opalescence	75
– The blue of the sky	60 (?)
Black body spectrum	64
Charged spheres (in a gas)	68
Radioactivity—Charges produced	62.5
– Helium engendered	64
– Radium lost	71
– Energy radiated	60

> Our wonder is aroused at the very remarkable agreement found between values derived from the consideration of such widely different phenomena. Seeing that not only is the same magnitude obtained by each method when the conditions under which it is applied are varied as much as possible, but that the numbers thus established also agree among themselves, without discrepancy, for all the methods employed, the real existence of the molecule is given a probability bordering on certainty. (Perrin 1916, 206–207; the question mark in the table is Perrin's)

One can hardly expect a clearer example of robustness reasoning. The analogous tables in *Brownian Movement and Molecular Reality* (Perrin 1910) as well as in the next English translation of *Atoms* (Perrin 1923), a translation of the 11th edition of *Les Atomes,* are very similar, though they do differ from each other in subtle ways: They sometimes cover different phenomena or give different values for *N* (under the same category). Indeed, we might anticipate such a progression in Perrin's work: With time his reasoning arguably improves by virtue of his dropping some phenomena and adding others and by employing various calculational and experimental corrections.

However, such diachronic variance is somewhat of a puzzle from the perspective of robustness. For example, if the earlier robustness argument in Perrin (1910) is found to be flawed because it cites illusory or irrelevant phenomena or makes faulty calculations, and if the later robustness argument in Perrin (1916) corrects these problems, what are we to make of the cogency of the earlier argument? Suppose that the convergent results in the earlier argument are still surprising to us (or to Perrin), despite the fact that we now think the results contain errors or make faulty assumptions. Should arguing robustly on the basis of the earlier results still be compelling to us, given that errors have been identified? If so, what are we to make of the cogency of robustness reasoning, if it can proceed on the basis on faulty results?

Of course we have suggested (in chapter 1) that robustness reasoners would want to make use of a 'minimal reliability requirement' whereby, reiterating Sober, the probability that an observation report is issued by an observational procedure (such as a report providing a value for Avogadro's number) is greater given the truth of this report than given its falsity. However, it is not easy to determine whether this condition is satisfied in the case of Perrin's research since, at the time Perrin is writing, one is unable to check how close either his earlier or his later assessments of Avogadro's number are to the *real* Avogadro's number. Moreover, even if we did determine that Perrin's earlier research is reliable enough (though less reliable than his later research), it is still unclear whether we really want to use this research for the purposes of a grand robustness argument involving the results from Perrin's both early and later work. This is because it is doubtful that the reliability of an observational procedure is enhanced by showing that it generates the same result as a different, but less reliable observational procedure. On the other hand, none of this progression in the quality of research forms much of an obstacle if one is utilizing what I have called 'reliable process reasoning', since it is precisely the goal to have an observational procedure that is maximally reliable from the perspective of the participant scientists. Such a goal motivated DAMA's preference for a model-independent approach to WIMP detection and motivated as well the emphasis microbiologists placed on empirically testing the assumptions underlying their experimental inquiries into mesosomes. Since a progression in the reliability of observational procedures is exactly what is sought, there is no need to bother with the question of what to do with less (though at least minimally) reliable, alternate observational approaches. These other approaches can be simply and safely ignored.

In any event, my plan is not to dwell on the challenge raised for robustness reasoning by the progression of Perrin's preferred observational methods. We will, for the sake of convenience, simply take as our primary and stable guide Perrin (1916), which is mostly reproduced verbatim in Perrin (1923) (though we make note of any important divergences). We also track for comparative reasons the discussion in Perrin (1910) and note here as well any important divergences between it and Perrin (1916). Finally, where relevant, we consider Perrin's views as expressed in his 1926 Nobel lecture. The result, I hope, is a dynamic picture of the kinds of phenomena Perrin cites for his purported robustness argument(s) with the goal of providing us with a comprehensive understanding of how Perrin thinks he justifies the tabulated values for Avogadro's number. In the end, we address the key question: What is the connection between the convergent values for Avogadro's number and the reality of atoms and molecules? It turns out that the answer isn't robustness after all.

THE VISCOSITY OF GASES

Our tack in examining Perrin's reasoning is to work our way down Perrin's table (as reproduced above), initially examining each line to see how Perrin justifies the values he provides for Avogadro's number. The first line of the table concerns the viscosity of gases for which N is given the value $62 \cdot 10^{22}$, and the justification for this value occurs in chapter 2 of Perrin (1916), section 46. This section occurs under a larger sectional heading, 'Molecular Free Paths', and Perrin's first task is to define the notion of a 'mean fee path.' Where we are considering a gas that is becoming mixed by means of diffusion, and in reflecting on how molecules in such a gas move by bouncing off one another, 'the *mean fee path* of a molecule . . . is the mean value of the path traversed in a straight line by a molecule between two successive impacts' (Perrin 1916, 74; Perrin's italics). Perrin notes (76) that one can calculate the mean free path using Maxwell's viscosity equation: Where ζ is the coefficient of viscosity, d is the gas density, G the mean molecular velocity and L the mean free path,

$$\zeta = G \cdot L \cdot d/3$$

As all the variables here, except for L, are measureable, one has a way to calculate L. From here, Perrin examines Clausius's relation between L and the diameters of the molecules in the gas. Roughly, the greater the diameter, the shorter the mean path; for simplicity, Clausius assumes that the molecules are spherical. Formally, where n is the number of molecules in cubic centimetre and D is the diameter of a molecule,

$$L = 1/\left(\pi \sqrt{2}\, n\, D^2\right)$$

(from Perrin 1910, 15). Now, at the time Perrin was writing, Avogadro's hypothesis had long been established: In Perrin's words, 'equal volumes of different gases, under the same conditions of temperature and pressure, contain equal numbers of molecules' (1916, 18). Of course, there is nothing in Avogadro's hypothesis that mentions a particular number of molecules—nor should it, because that number varies with the temperature, volume and pressure. So Perrin sets up a convention (analogous to conventions currently used): He defines a 'gramme molecule' (what we now call a 'mole') as follows:

> The gramme molecule of a body is the mass of it in the gaseous state that occupies the same volume as 32 grammes of oxygen at the same temperature and pressure (i.e., very nearly 22,400 c.c. under 'normal' conditions). (1916, 26)

Let us look at Perrin's convention this way: 32 grams of oxygen gas at a certain designated temperature and pressure occupy a volume v. In this volume, the number of oxygen molecules is called Avogadro's number, N. For any other kind of gas under the same conditions, if the gas contains Avogadro's number of molecules, then the gas will occupy the same volume, and we can be said to have a gramme molecule of this gas. So suppose we have a gramme molecule of a gas, containing N molecules and occupying a volume v in cubic centimetres; then the number of molecules in a cubic centimetre $n = N/v$. We can now substitute N/v for n in Clausius's equation:

$$\text{(C)} \quad \pi N D^2 = v / (L\sqrt{2})$$

(from Perrin 1916, 78). In this equation there are two unknowns, N and D. The next step is to find a formula that relates these two variables.

Perrin's first attempt at this formula considers N spherical molecules, each of diameter D, resting as though they were in a pile of shot; he notes that the volume occupied by such spheres, $\pi ND^3/6$, is less than the entire volume of the pile by at least 25% (Perrin 1910, 15, and Perrin 1916, 79). This inequality, in turn, combined with Clausius's equation (C), allows Perrin to set a lower limit to N (and an upper limit to D). The value at which he arrives, where we are considering mercury gas (which is monatomic, so its molecules are approximately spherical) is $N > 44 \cdot 10^{22}$ (Perrin 1916, 79; Perrin 1910 cites the value $N > 45 \cdot 10^{22}$). In Perrin (1910), he records his attempt at a similar calculation with oxygen gas (he neglects to mention this attempt in 1916), giving a value of $N > 9 \cdot 10^{22}$. This value he found to be far too low; he describes the mercury value as 'higher and therefore more useful' (16). In Perrin (1910), he also performs a calculation that serves to determine an upper limit to N using Clausius's and Mossotti's theory of dialectrics (16–17). By this means, using the case of argon, he arrives at the value $N < 200 \cdot 10^{22}$. The inequalities, $45 \cdot 10^{22} < N < 200 \cdot 10^{22}$, are recorded by Perrin in his summarizing table at the end of Perrin (1910) (an analogous table to the one we cited above). As such, they form part of Perrin's (1910) 'proof of molecular reality' (90).

In *Atoms* (Perrin 1916 and Perrin 1923), Perrin completely omits these inequalities in his table and completely omits the discussion of an upper limit to N. As regards the calculation of a lower limit using mercury gas, he complains that it leads to values 'too high for the diameter D and too low for Avogadro's number N' (1916, 79). To some degree, then, Perrin is being selective with his data, and one might legitimately suggest that if one plans to use robustness reasoning to determine whether observational procedures are reliable, one should not be antecedently selective about the observed results that form the basis of a robustness argument.

This is excusable if a rationale can be given for why results are omitted, and one is provided in Perrin (1910), though not in Perrin (1916). In essence, Perrin is concerned that the pile of shot method is not very reliable since 'we only know how to evaluate roughly the true volume of *n* molecules which occupy the unit volume of gas' (1910, 17).

Recall that the challenge in using Clausius's mean free path equation (C), if we want to provide a determination of *N*, is to functionally relate *N* and *D*, and Perrin notes that 'a more delicate analysis' (1910, 17) can be found in the work of van der Waals. Van der Waals's equation is a generalization of the ideal gas law that takes into account the non-negligible volumes of gas molecules (symbolized as *B* by Perrin, 1916) as well as the forces of cohesion between these molecules (symbolized by Perrin as *a*). As *B* and *a* in any observational application of van der Waals's equation are the only two unknowns, two separate applications of the equation can be used to solve for each of these variables. Thus, whereas before we had only a vague estimate for $\pi ND^3/6$, we now have

$$\pi ND^3/6 = B$$

with only *N* and *D* unknown, which allows us to solve for each unknown given (C). Along these lines, Perrin works out values for *N*, deriving '$40 \cdot 10^{22}$ for oxygen, $45 \cdot 10^{22}$ for nitrogen, [and] $50 \cdot 10^{22}$ for carbon monoxide, a degree of concordance', he says, 'sufficiently remarkable' (1916, 81). One might expect Perrin to argue robustly here for the accuracy of these values, but he rejects these values because molecules of oxygen, nitrogen and carbon dioxide are not spherical and so, he is concerned, are 'not best suited to the calculation'. Argon, by comparison, 'can give a trustworthy result' (81), leading to the value $62 \cdot 10^{22}$. This result is then dutifully recorded in Perrin's (1916) summarizing table. In an apparent typographical error, he records $60 \cdot 10^{22}$ in the parallel table in Perrin (1910).

In Perrin (1923), by comparison, Perrin appends an '(?)' to this value in his table, indicating a growing uncertainty on his part about this calculation of *N*. Indeed, in all three sources, (Perrin, 1910, Perrin, 1916 and Perrin, 1923), he notes that this calculation of *N* has a large error—40% in Perrin (1910, 48) and 30% in Perrin (1916) and Perrin (1923)—'owing to the approximations made in the calculations that lead to the Clausius–Maxwell and van der Waals equations' (1916, 82). This is a significant source of error and one might justifiably wonder whether this result should be included at all. In this respect, Perrin (1910), Perrin (1916) and Perrin (1923) differ in their assessments. Acknowledging this large error, Perrin (1910) comments, 'by methods completely different we proceed to consider similar results for which the determination can be made with greater accuracy' (18); in other words, he seems ready to accept his calculation of *N* if it is found to cohere with results more reliably produced. On the other hand, Perrin (1916) and Perrin (1923) comments, 'if by entirely independent routes we are led to the same values for the molecular magnitudes, we shall certainly find our faith in the theory considerably strengthened' (82, both editions), which seems to be as pure an expression of robustness reasoning as one can find. It remains to be seen if Perrin succeeds in carrying out this promise of a robustness argument in subsequent chapters; to foretell our results, the story turns out somewhat differently.

Before proceeding, an important issue we need to consider is whether the results Perrin has retrieved so far, despite their uncertainty, are nevertheless significant. One might claim here, as Perrin seems to do, that even though we have a 30% chance of error (where $N = 62 \cdot 10^{22}$, $37 \cdot 10^{22} < N < 80 \cdot 10^{22}$), we still have a surprising result concerning at least the order of magnitude of *N*. That is, we at least know that *N* is in the 10^{22} range. Isn't this order of magnitude result significant? And isn't it guaranteed by a robustness argument in which values for *N* within this error range are generated using mercury gas with a pile of shot calculation, as well as with oxygen, nitrogen, carbon monoxide and argon using the van der Waals calculation? Let us call this the 'order of magnitude' robustness argument for the determination of Avogadro's number—different lines of evidence have led to a determination of *N* 'within an order of magnitude', leading us to conclude that the value of *N* must be within this range. Surely, one might suggest, this is an effective argument. But if it is, it is somewhat of a mystery why Perrin continues to provide further, different approaches to determining *N*. As we shall see, the other determinations of *N* that Perrin provides using other, different routes are hardly more precise, if our focus is solely on orders of magnitude. (This becomes especially obvious once we consider that the current, best estimate for Avogadro's number is 60.22 141 79 $\cdot 10^{22}$, plus or minus 0.00 000 30 $\cdot$ 10^{22}; see Mohr et al. 2008). If it's the order of magnitude that we're after, two or three independent determinations should be sufficient to warrant surprise at a convergence of results. So why does Perrin think we need 13 such determinations? (As I shall suggest later on, one of the characteristic weaknesses of robustness reasoning is that it lacks specific guidelines on how many independently generated observed results are needed for a robustness argument to be effective.) Finally, if an order of magnitude result is all he's looking for, why would Perrin bother with a level of precision better than 30%?

There are further questions one might ask regarding the significance of a robust, order of magnitude result for Avogadro's number. One question focuses on how close the numbers $37 \cdot 10^{22}$ and $80 \cdot 10^{22}$ actually are, for from one perspective they are apart by $43 \cdot 10^{22}$, which is a very large number, an error practically as large as the estimate of *N* itself. Still, one might point out that having values of *N* all in the 10^{22} range is still significant enough. Similarly, one might say that the numbers 3 and 8 are close too, since they are both in the 10^0 range. But surely the matter of the closeness of numerical estimates is highly context dependent. For example, the numbers 3 and 8 are very close if we're asking about someone's yearly income in dollars but not close at all if we're considering a hockey score. Put another way, suppose one were to ask, 'What was your income last year?', and the response was, 'In the 10^0 range'—that would be an informative response. However, if one were to ask, 'How many goals did the hockey team score last night', and the response was, 'In the 10^0 range'—that would not be informative at all.

So what about an estimate of Avogadro's number as 'in the 10^{22} range'? Is this estimate informative? This may not be a question we can easily answer since it depends, as with incomes and hockey scores, on the context. That is, if the context allows for a potentially large range of possible values, as with incomes, then we've learned something significant with 'in the 10^{22} range'. But then, by analogy with hockey scores, it may be that the constitution of physical matter makes it is impossible for the number of atoms or molecules in a mole of gas at standard temperature and pressure to have an order of magnitude other than 10^{22}, a fact we would more fully appreciate if we understood better the atomic nature of matter (just as the limited range of hockey scores is comprehensible once we understand the game of hockey). To consider a different, sporting analogy, suppose one asks how many people there are in a football stadium on game day, and the answer is, 'In the 10^4 range'. Given that the stadium sits $7 \cdot 10^4$ people, and football enjoys a fair amount of popularity in the area, such an answer says practically nothing—even if one devises ingenious ways to robustly confirm this result, such as through visual density measurements from aircraft above the stadium, concession stand receipts, counting the cars in the parking lot and so on. A curiosity with Avogadro's number, however, is its enormous size, which for that reason makes it seems like an informative figure (just as, with football on game day, the neophyte fan might be shocked to learn that varsity draws 'tens of thousands of fans'). Along these lines, some authors like to put the vastness of Avogadro's number in perspective by using an analogy. For example, as Wisniak (2000) notes, 'An Avogadro's number of standard soft drink cans would cover the surface of the earth to a depth of over 200 miles' (267). This is an impressive picture, but the analogy may be misleading. We can imagine depths ranging from one can deep up to 200 miles of cans deep—nothing physically, so far as we can tell, precludes any value in this range. But atomic reality may be much different than this. It may just not be physically possible to have values of N ranging from the 100 range to anything less than 10^{22} or anything more than 10^{22}. If so, robust data showing that N has a value in the 10^{22} range, given that one is aware of such this impossibility, would not be terribly informative.

At this stage the proponent of the order of magnitude robustness argument may suggest that the presence of such physical impossibilities is irrelevant to the intentions of the argument. Rather, the argument is meant to impress in a case in which we don't know in advance, one way or another, what the order of magnitude of Avogadro's number is (or must be), and, as it happens, Perrin surprisingly finds a convergence around 10^{22} by means of different, independent routes in the absence of a prior knowledge of this order of magnitude. For comparison, consider again the analogy with attendance at a football stadium. If one already has a fairly good assurance that game day attendance is in the 10^4 range, an assurance gained perhaps by reflection on the size of the stadium and an awareness of the normal popularity of the sport in the region, it follows once again that devising ingenious ways to robustly confirm this result shows practically nothing. It is knowledge of the order of magnitude that we already have, and such robust results, if they don't improve on precision, would simply be redundant. Now it turns out that this was the situation with Avogadro's number at the time Perrin was writing his books, both *Brownian Movement and Molecular Reality* and *Atoms*; at that time, there was fairly strong assurance that Avogadro's number was indeed in the 10^{22} range, as Perrin himself acknowledges. For instance, Perrin (1910, 76) and Perrin (1916, 128) both cite Einstein's (1905) value for *N*, $40 \cdot 10^{22}$, and in a footnote in Perrin (1916) to his discussion of Einstein's result, Perrin mentions Theodor Svedberg's (1909) value of $66 \cdot 10^{22}$. Perrin (1910, 89–90) also mentions previous values of *N* generated by a consideration of dark radiation: Lorentz's value of $77 \cdot 10^{22}$ and Planck's value of $61 \cdot 10^{22}$. In fact, as John Murrell (2001, 1318) points out, an estimate of *N* was available as early as 1865 in the work of Josef Loschmidt, who calculated the number of molecules per cubic centimeter of gas at standard temperature and pressure, instead of (as with Perrin) per mole (or gramme molecule). Murrell asserts that Perrin had calculated Loschmidt's number to be $2.8 \cdot 10^{19}$, quite close to the currently accepted value of $2.7 \cdot 10^{19}$ (2001, 1320). For his part, Loschmidt in 1865 arrived by means of an erroneous calculation at the value of $8.66 \cdot 10^{17}$ for his namesake number. Subsequently, a corrected calculation was performed by J. C. Maxwell in 1873 leading to a value of $1.9 \cdot 10^{19}$, which is clearly a result that when converted according to Perrin's convention would generate a value for Avogadro's number of the right order of magnitude (Murrell 2001, 1319). Here we should be careful not to underestimate the importance of Loschmidt's contribution. Murrell comments that 'in the German literature one often finds Avogadro's constant referred to as Loschmidt's number per gram molecule' (1318, footnote 7). This observation is echoed by Virgo (1933) who remarks,

> The first actual estimate of the number of molecules in one cubic centimetre of a gas under standard conditions was made in 1865 by Loschmidt, and from this the number of molecules (atoms) in a gram molecule (atom) was later evaluated. From the quantitative view-point it thus seems preferable to speak of "Loschmidt's

> number per gram-molecule (atom)," and of "Loschmidt's number per cubic centimetre," as is almost invariably done in the German scientific literature. (634)

The significance of Maxwell's contribution should also not be downplayed. As Charles Galton Darwin points out in his 1956 Rutherford Memorial Lecture, 'the first estimate of Avogadro's number is due to Maxwell himself'. Here, Darwin is well aware of the two conventions regarding the definition of Avogadro's number, Loschmidt's and Perrin's, commenting that 'it has been found convenient to define [Avogadro's number] not in terms of the number of atoms in a cubic centimeter of gas, but as a number in a gram-molecule of any substance' (1956, 287). He then cites the value of 'Loschmidt's number' attributed above to Maxwell (i.e., $1.9 \cdot 10^{19}$, though he calls it 'Avogadro's number') and remarks that

> [Maxwell's] result may not seem very accurate, but when consideration is given to some of the rather doubtful details, I think the answer might easily have come out much further from the truth. (287)

So Maxwell's result, it seems, had at least the merit of having the right order of magnitude; and this result, as Darwin continues, was subsequently confirmed by Rayleigh's molecular explanation for the blueness of the sky that produced a value for Avogadro's number 'that entirely confirmed Maxwell's [value], but did not narrow the limits of the accuracy to which it was known' (287).

Let us acknowledge, then, that the scientific community for whom Perrin was writing was well aware of what order of magnitude should be expected from a determination of Avogadro's number. It follows that Perrin's presumed order of magnitude robustness argument was not for his contemporaries—or, at least, should not be for us—very informative, here taking a subjective perspective. Objectively, on the other hand, the matter is somewhat indeterminate given, as I have suggested, a lack of awareness of what values of N are physically possible. So overall my submission is that we should view the order of magnitude argument as somewhat limited in regards to what it can tell us, both scientifically and historically, and we should not overplay its significance. Even more to the point, it is clear that Perrin seeks far greater precision in a determination of Avogadro's number than simply an order of magnitude.

Let us now turn to the next line in Perrin's table, the first of three lines motivated by the phenomenon of Brownian movement.

BROWNIAN MOVEMENT: VERTICAL DISTRIBUTIONS IN EMULSIONS

Small particles suspended in a fluid, similar to dust particles seen in sunbeams, exhibit an endless, seemingly random movement called 'Brownian motion', named after the Scottish microscopist who observed it in 1827. Following the work of Louis Georges Gouy, Perrin notes that the particles subject to Brownian motion are unusual in that their movements are completely independent of one another (Perrin 1910, 5, and Perrin 1916, 84) and thus are not caused by currents in the sustaining fluid. In addition, Brownian motion falsifies a deterministic reading of the Second Law of Thermodynamics (called 'Carnot's Principle' by Perrin) prohibiting the transformation of heat into work—for example, a Brownian particle might spontaneously rise upwards against gravity without the expenditure of energy (Perrin 1910, 6–7, and Perrin 1916, 86–87). To explain these unusual characteristics, Gouy hypothesized that Brownian particles are caused by the motion of molecules (Perrin 1910, 7, and Perrin 1916, 88–89). Though Perrin is impressed with this hypothesis, he asserts that we need to put it to a 'definite experimental test that will enable us to verify the molecular hypothesis as a whole' (1916, 89).

Perrin's ingenious approach to putting the molecular hypothesis to a test is the basis for his receipt of the Nobel Prize in 1926. To begin, he cites received knowledge about the distribution of gas molecules in vertical columns, according to which a gas higher in the column will be more rarefied than the portion of gas lower in the column. He then calculates precisely how the pressure of a gas at a lower elevation p is related to the pressure of gas at a higher elevation p': where M is the mass of a gram molecule of the gas, g is the acceleration due to gravity, h is the difference in elevation, R is the gas constant and T the absolute temperature,

$$\text{(P)} \quad p' = p\,(1 - ((M \cdot g \cdot h)/RT))$$

We see, then, that for every distance h we ascend, the pressure is reduced by a common factor $(1 - ((M \cdot g \cdot h)/RT))$, which means that the pressure exhibits an exponential progression. Also, the common factor is found to directly vary with M, so that for larger molecular sizes the rarefaction at higher altitudes proceeds more quickly. Finally, since the pressure of a volume of gas is proportional to the number of molecules in this volume, we will find a similar geometric progression when we compare the number of molecules at a lower elevation to the number at a higher elevation.

At this stage, Perrin (1916) asks us to consider an analogous substance to a gas, that is, a uniform emulsion (also called a colloid). An emulsion contains particles that are suspended in a fluid and that move about in Brownian fashion; it is uniform if its constituent particles are the same size. An emulsion, if it is bounded by a semipermeable membrane, will exert a pressure on this membrane that, by van't Hoff's law, is analogous to the pressure exerted by a gas on the walls of a container. Specifically, this

> osmotic pressure [will be] equal to the pressure that would be developed in the same volume by a gaseous substance containing the same number of gramme molecules (39),

and so, by Avogadro's hypothesis,

> either as a gas or in solution, the same numbers of any kind of molecules whatever, enclosed in the same volume at the same temperature, exert the same pressure on the walls that confine. (39)

In other words, gases and emulsions form a continuum in term of how they express the phenomenon of pressure: Emulsions, in effect, simply contain large uniform particles whereas a gas contains much smaller particles (i.e., molecules). Thus, for the equation (P) above relating the pressures exerted by a gas at different elevations, there is an analogous equation relating the osmotic pressures exerted by an emulsion at different heights. Where we are considering the numbers of particles (as opposed to the osmotic pressure) n in an emulsion at a lower elevation as compared to the number n' at a higher elevation, and where we take into account the buoyancy of the liquid constituting the emulsion by means of the factor $(1 - d/D)$, with d standing for the density of the liquid and D the density of the emulsive particles, with the gramme molecular weight of these particles signified by $N \cdot m$ (m is the mass of each particle, assumed to be uniform in size),

$$n' = n\,(1 - (N \cdot m \cdot g \cdot h\,(1 - d/D)/RT))$$

The significance of this vertical distribution equation cannot be underestimated: If we can count the numbers of emulsive particles at different heights, we have enough information to directly calculate N, Avogadro's number.

For this calculation to work, one needs to prepare suitable emulsions whose particulate matter is uniform in size (to complete the analogy to uniformly sized gas molecules). Perrin successfully used two sorts of emulsions, one with gamboge and the other with mastic, and describes in detail in Perrin (1910, 27–29) and Perrin (1916, 94–95) how he prepared these emulsions by means of fractional centrifugation. With the emulsions at hand, in order to apply the vertical distribution equation, two quantities need to be worked out: the mass m as well as the density D of the emulsive particles. In Perrin's (1916) determinations of these quantities, he suggests that he arrives at these quantities by reasoning on the basis of 'concordant' observations (that is, using robustness reasoning). Supposedly, then, robustness plays a central role for Perrin not only in his overall argument for the accuracy of his determination of Avogadro's number (using his table) but also in his more local arguments for the values of certain key observed quantities. Unfortunately, his determinations of m and D in Perrin (1916; identically reproduced in Perrin 1923) are a source of some confusion, particularly if we take them to exemplify robustness reasoning.

Take for instance his discussion of how one works out the density of the emulsive granules. Perrin says,

> I have determined this in three different ways:
>
> (a) By the specific gravity bottle method, as for an ordinary insoluble powder. The masses of water and emulsion that fill the same bottle are measured; then, by desiccation in the oven, the mass of resin suspended in the emulsion is determined. Drying in this way at 110° C. gives a viscous liquid, that undergoes no further loss in weight in the oven and which solidifies at the ordinary temperature into a transparent yellow glass-like substance.
> (b) By determining the density of this glassy substance, which is probably identical with the material of the grains. This is most readily done by placing a few fragments of it in water, to which is added sufficient potassium bromide to cause the fragments to remain suspended without rising or sinking in the solution. The density of the latter can then be determined.
> (c) By adding potassium bromide to the emulsion until on energetic centrifuging the grains neither rise nor sink and then determining the density of the liquid obtained.
>
> The three methods give concordant results. (95)

What is puzzling is that the two methods, (a) and (b), are viewed as *one* method in Perrin (1910) (and also viewed as one method in Nye 1972, 106) and that Perrin (1910) presents an entirely different, fourth method for determining the density of granules that is said by him to be 'perhaps more certain' (29), though it is entirely omitted in Perrin (1916). To further complicate matters, in his 1926 Nobel lecture Perrin asserts that

> there is no difficulty in determining the density of the glass constituting the spherules (several processes: the most correct consists in

> suspending the grains in a solution which is just so dense that the centrifuging cannot separate the grains) (149),

thus suggesting that method (c) is in fact the best method, contrary to Perrin (1910), and without any consideration of the special value of concordant results. In other words, Perrin's (1916) alleged allegiance to a form of robustness reasoning in determining the density of emulsive particles is hermeneutically problematic if we take into account Perrin (1910) and Perrin (1926).

Perrin's calculations of mass suffer from a similar difficulty in interpretation as well. Just as with his determinations of particle density, Perrin describes his determination of particle mass as involving three differing methods that converge in their results. Two of the methods involve direct determinations of the radius of emulsive granules, determinations that when combined with a previous knowledge of granule density gives us the mass of the granules. With the first method (Perrin 1910, 38, and Perrin 1916, 96–97), a dilute emulsion is allowed to dry with the result that some of the granules line up in rows only one granule deep. The length of these rows is much easier to measure than individual granules, and by simply counting the grains in a row one arrives at the radius of a granule. The second method (Perrin 1910, 34–40, Perrin 1916, 97–99; see also Nye 1972, 108–109) involves the use of Stoke's law, which relates the velocity of a spherical particle falling through an atmosphere with a particular viscosity. Applied to the case of a uniform emulsion, all the variables in Stoke's law can be measured, except for the radius of particles, which can then be calculated. The third method involves what Perrin calls a 'direct weighing of the grains' (1916, 97): An emulsion is made slightly acidic with the result that the granules attach themselves to the walls of the container, allowing them to be counted. With a prior knowledge of the concentration of the emulsion the mass of the particles can be determined, and from here we can arrive at their radii. As each of these methods arrives at concordant results for the radius of a granule, we seem to have a solid justification for this radius. Indeed, Perrin says, 'It is possible, on account of the smallness of the grains, to place confidence only in results obtained by several different methods' (1916, 96). However, a closer look at Perrin's thinking reveals that the situation is more complicated.

Of particular concern is the justification of an application of Stoke's law to the case of an emulsion. As Perrin notes, Stoke's law had originally been formulated to apply to much larger particles, such as water droplets or bits of dust, and Jacques Duclaux, for instance, had expressed reservations about the propriety of extending Stoke's law to emulsive granules. For Perrin, the agreement he finds with the results derived from Stoke's law and the results generated by the other two methods answers Duclaux's doubts. He says, 'The concordance of the preceding measurements will dispel these doubts.... The preceding experiments show that this law is valid in the domain of microscopic quantities' (1910, 40). Also, Perrin (1916) remarks: 'It cannot now be doubted, in the face of the concordant results given above, that in spite of the Brownian movement the extension of [Stoke's] law is legitimate' (99). But what is being described here is not a form of robustness reasoning but a case of calibration. Perrin is suggesting that since Stoke's law generates the same results as two other, more reliable methods, its own reliability is assured. This assessment is echoed in Nye (1972) who describes the direct weighing of the grains method as constituting a sort of 'control experiment' for the application of Stoke's law to emulsions (109). The legitimacy of viewing Perrin's reasoning in this way—as reading 'concordant' results as 'calibrated' results—should be apparent when we consider that Perrin did not consider the other two methods to be nearly as controversial. He applauds the validity of the direct weighing method for explicitly avoiding the application of Stoke's law (1910, 37). Moreover, in his tabulation of the results of measuring the radii of gamboge granules (1910, 39), the line with greatest accuracy involves a straightforward comparison of the direct weighing method and Stoke's method, which would mean that the accuracy of the results rests directly on the former method, given Perrin's apprehension with the Stoke's law approach. Finally, when we look again at Perrin's Nobel Prize lecture, all direct references to Stoke's law are omitted, the celebrated 'concordance of results' is ignored and the only observational justification cited involves the first method, the measurement of dried rows of gamboge grains.

Having established how one should go about working out the values of the variables to his vertical distribution equation, Perrin conducts a number of experiments to see, first, whether the values of n' and n exhibit a geometrical progression as one moves to higher elevations in the emulsion (they do, vindicating the analogy to molecules in a gas) and, second, to calculate the values of N in each experiment. Here, once more, Perrin's stated strategy is to employ robustness reasoning: He uses varied experiments, such as using different sizes of emulsive grains (from .14 to 6 microns), different intergranular liquids (water, sugary water, glycerol and water), different temperatures for the intergranular liquid (–9° C to 60° C) and different kinds of emulsive grains (gamboge and mastic), and with all these methods arrives at a value of N in which $65 \cdot 10^{22} < N < 72 \cdot 10^{22}$ (Perrin 1910, 44–46, Perrin 1916, 104–105, Perrin 1926, 150). On the basis of these experiments, he asserts that he has 'decisive proof' of the existence of molecules (1916, 104). What is the nature of this proof?

In Perrin (1926), he takes the surprising fact that the values of n' and n exhibit a geometrical progression at all as justification for the molecular hypothesis:

> The observations and the countings...prove that the laws of ideal gases apply to dilute emulsions. This generalization was predicted as a consequence of the molecular hypothesis by such simple reasoning that its verification definitely constitutes a very strong argument in favour of the existence of molecules. (150)

A similar consideration motivates Perrin (1916):

> even if no other information were available as to the molecular magnitudes, such constant results would justify the very suggestive hypotheses that have guided us, and we should certainly accept as extremely probable the values obtained with such concordance for the masses of the molecules and atoms. (105)

That is, the value of N—whatever it is—must be constant in order for the analogy between for the behaviour of gases and uniform emulsions to succeed, for the density of (the molecules in) a gas at different heights exhibits a geometrical progression, and unless N is found to be constant, one would not anticipate seeing such a progression with emulsions. This is a fairly straightforward analogical argument, though one might hesitate to call it a strong argument. To begin with, gases at different heights might exhibit densities that express a geometrical progression, but this may not be because they contain molecules—rather they might contain uniformly sized gas particles of another sort. Moreover, though the results with the emulsions are constant under varying conditions, the conditions are not *that* varied: We are dealing only with gamboge and mastic under a relatively narrow temperature range with a somewhat narrow range of grain sizes, and our conclusion purports to encompass the whole range of physical matter that is possibly constituted by atoms and molecules.In fact, Perrin (1916) has a stronger argument he wishes to propose, one that takes into account the specific values he calculates for 'the molecular magnitudes'. He begins by noting that, with his viscosity measurements, he had retrieved a value for $N = 62 \cdot 10^{22}$, a number he takes to be surprisingly close to the range he derived with the vertical distribution measurements. 'Such decisive agreement', he submits, 'can leave no doubt as to the origin of the Brownian movement', for

> it cannot be supposed that, out of the enormous number of values a priori possible [for the emulsion measurements], values so near to the predicted number [the viscosity number] have been obtained by chance for every emulsion and under the most varied experimental conditions. (105)

Almost exactly the same wording is used in Perrin (1910, 46). The key word here is 'predict': On the basis the viscosity measurements, Perrin makes a novel prediction as regards the emulsion measurements—'novel' in that, a priori, he thinks, most any value for N had been possible with the emulsion measurements prior to the viscosity measurements. But, if Perrin's argument is based on the epistemic merit of novel prediction, that is a very different issue from the question of robustness. Recall that Perrin's presumed, overall robustness argument, the details of which are summarized in his table, draws from a variety of other methods, not just viscosity and emulsion measurements. But here, in discussing the emulsion results, he is asserting that he has found 'decisive' proof for molecular reality, one that leaves us with 'no doubt'. So is there much need for the other methods he describes? There may not be, if he feels comfortable with the reliability of his experimental determination of N using emulsions and if he is allied to a methodology that places an emphasis on the epistemic value of novel predictions.

But there is a reason to resist reading Perrin as a predictivist, which also serves as a reason to resist reading him as an advocate of robustness reasoning. The problem is that the viscosity measurements are viewed by Perrin as involving a significant amount of error: 40% in Perrin (1910) and 30% in Perrin (1916), as we saw above, and 100% in Perrin (1926, 143). Moreover, as Perrin emphasizes, this error value cannot be reduced (Perrin 1910, 48, and Perrin 1916, 107), since the viscosity measurements ineliminably depend on certain dubious assumptions. So it is hard to see what epistemic merit we can attach to these measurements (unless we are simply looking at order of magnitude considerations, as Perrin 1926 suggests, 143); as such, they form a weak basis on which to ground either a novel prediction or a robustness argument. Another consideration here is that, despite providing a number of measurements of N using varying methods, Perrin is also concerned about generating a best value for N using the most ideal assumptions. For Perrin (1910), this best value was derived using gamboge grains with a radius of .212 microns, leading to a value for $N = 70.5 \cdot 10^{22}$. This value subsequently occurs in the table at the end of the book. For Perrin (1916), the best value was derived using grains with a radius of .367 microns, leading to a value of $N = 68.2 \cdot 10^{22}$—we saw this value in the table at the beginning of this chapter. As we shall see, these best values are critical for Perrin's subsequent arguments for the reality of molecules, and their merit lies in the reliability of the method by which they were generated, not in the fact that they were predicted using viscosity measurements (these methods were imprecise and, Perrin acknowledges, error-ridden), nor in the fact that they were generated using diverse methods (strictly speaking, other methods yielded different, precise values).

BROWNIAN MOVEMENT: DISPLACEMENT, ROTATION AND DIFFUSION OF BROWNIAN PARTICLES

Working again with emulsions, Perrin considers in the next line in the table the laws governing the displacement of emulsive particles (as distinct from considering the vertical distribution of emulsive particles generated by these laws). If we assume that these displacements are completely irregular, and if we further assume that we can treat analogically the diffusion of grains in an emulsion as though it behaves like the diffusion of molecules in a solution, then famous work by Einstein (1905) suggests the following mathematical treatment. First, the emulsive particles diffuse (just as molecules in a solution do) in accordance with Maxwell's distribution law for molecular speeds (Perrin 1910, 52, and Perrin 1916, 117). If we assume further that the emulsion is in equilibrium, with the upwards diffusion of particles equally balanced by the fall of particles due to gravity, then Einstein supposes that this fall can be described by means of Stoke's law (Perrin 1910, 53, and Perrin 1916, 113). Of course, the use of Stoke's law in this context is problematic, and Nye (1972) notes that some of Perrin's contemporaries, such as Victor Henri, expressed skepticism about Einstein's calculations for just this reason (126). Perrin, however, believes he has put Stoke's law on firm footing (as we saw above), and so he is supportive of the following relation derived by Einstein: Where x^2 is the mean square of the projection of the displacement of an emulsive particle along an arbitrary axis, t is the time elapsed, R is the gas constant, T is the absolute temperature, a is the radius of an emulsive particle and ξ is the viscosity of the fluid,

$$\text{(E)} \quad x^2/t = (R \cdot T) \,/\, (N \cdot 3\,\pi\, a\, \xi)$$

(Perrin 1910, 53, Perrin 1916, 113). Since all of the variables in (E) can be measured, except for Avogadro's number N, we presumably have a way to determine N. From here, we might expect Perrin to argue robustly as follows: Given that N derived in this way coheres with N derived earlier from the vertical distribution (and viscosity) measurements, one has the basis to argue for the accuracy of N so derived and from here argue in support of the molecular hypothesis (in a way, however, that is never made entirely clear).

But Perrin (1910) and Perrin (1916) argue in a very different way when one looks at the details of his discussions, a fact that is concealed if one examines exclusively his summarizing discussions pertaining to the tables found at the end of his monographs. The main point to make in this regard is that Perrin views his experiments on gamboge and mastic emulsions as confirming Einstein's equation (E); he does not regard himself as using (E) to simply calculate values of *N*. For example, Perrin (1910, 54–59) examines earlier attempts at confirming (E), citing 'the presumption of a partial verification' (56) by Franz Exner, followed by a purported confirmation by Theodor Svedberg that Perrin considers flawed. He comments, 'The obvious conclusion from the experiments of Svedberg [is], contrary to what he says, that the formula of Einstein is certainly false' (57). He also considers the experiments of Victor Henri, which lead to results Perrin views as 'completely irreconcilable with the theory of Einstein' (1910, 58). Similarly, Perrin (1916) mentions Max Seddig's 'partial verification' (120) and Victor Henri's (kinematographic) experiment 'in which for the first time precision was possible [and that] led to results distinctly unfavourable to Einstein's theory' (121). By 1909, the tide had turned in physicists' minds away from asserting the validity of Einstein's equation, a fact that Perrin (1910) ascribed to a regrettable short-sightedness. He comments that these negative results

> produced, among the French physicists who closely follow these questions, a current of opinion which struck me very forcibly as proving how limited, at bottom, is the belief we accord to theories, and to what a point we see in them instruments of discovery rather than of veritable *demonstrations*. (58; Perrin's italics)

He makes comparable remarks in Perrin (1916):

> I have been very much struck by the readiness with which at that time it was assumed that the theory [of Einstein] rested upon some unsupported hypothesis. I am convinced by this of how limited at bottom is our faith in theories. (121–122)

These comments are significant in that they reveal a certain theory-centeredness in Perrin's mind, a resistance to what is being learned empirically. But this does not stop him from attempting to put Einstein's formula on firm empirical footing, which he does in both Perrin (1910) and Perrin (1916).

To this end, Perrin's first task in Perrin (1910) is to draw attention to errors in both Svedberg's and Henri's experimental work (56–59). Doing that is important, but the main task for Perrin is to describe his own improved experimental methods, methods that generate more accurate values for the variables in (E) and that, accordingly, produce a more accurate value for *N*. Just as with his vertical distribution experiments, Perrin generates results that involve a variation of certain experimental parameters. He uses emulsive grains of different sizes, different kinds of intergranular fluids (such as sugar solution, urea solution and glycerine) and different sorts of grains (gamboge and mastic). Yet in both Perrin (1910) and Perrin (1916), he expresses a clear preference for certain particular assignments of these values. In Perrin (1910), gamboge grains of .212 microns 'served for [his] most exact determination of *N*' (60), just as it did for his vertical distribution calculations. Using grains of this size, he produces a value for $N = 70 \cdot 10^{22}$ which, he notes, 'is practically identical with that found by the completely different [vertical distribution] method [i.e., $70.5 \cdot 10^{22}$]' (61). Averaging in the results for mastic produces $N = 71.5 \cdot 10^{22}$ (the value he includes in the table in Perrin 1910), again agreeing with the vertical distribution result. Having produced these results, he feels comfortable in asserting that Einstein's formula is confirmed. But to say that this formula is confirmed is very puzzling if our purported goal is to arrive at a better confirmed value for *N* on the basis of a convergence of results. With robustness, the procedural correctness of one of the ways of generating the result should not be at issue; we are to assume the relative reliability of each of these ways and then argue for the accuracy of a convergent, observed result. But here with Perrin the goal, rather, is to argue for the accuracy of Einstein's formula by showing that it generates the same result as the one arrived at with the distribution experiment: In effect, we are *calibrating* Einstein's method by exhibiting its consistency with another approach whose reliability is not subject to scrutiny.

The same style of argumentation occurs in Perrin (1916). In tabulating his retrieved values for *N* on the basis of displacement measurements using emulsions (Perrin 1916, 123), he generates the range, $55 \cdot 10^{22} < N < 80 \cdot 10^{22}$, which is in fact not much better than the range generated through the viscosity of gases calculation. Still, Perrin notes that the average value for this range ('in the neighbourhood of 70 [$\cdot\ 10^{22}$]') is close enough to the value generated in the vertical distribution experiment to '[prove] the rigorous accuracy of Einstein's formula' (123). He also says that it also confirms 'in a striking manner... the molecular theory' (123), though Perrin never quite explains how this is so. Perrin, however, does not rest content with the range of values he has produced. He goes further and specifies what he claims to be 'the most accurate measurements', measurements involving gamboge grains with a radius of .367 microns. After explaining why he regards these measurements as the most accurate, he notes that the resultant calculated value of N is $68.8 \cdot 10^{22}$ (the value that is recorded in his table), quite close to the value of $68.3 \cdot 10^{22}$ produced in the distribution experiments. Not only then does Perrin not seem to be arguing robustly for the accuracy of his values for N (he is, again, calibrating his displacement measurements using his preferred distribution results). He is, rather, using a form of reliable process reasoning to argue for the accuracy of his displacement results by using a form of reasoning that starts with the assumed reliability of a procedure that generates these (displacement) results (that procedure using gamboge grains with a radius of .367 microns) and then accepts as most accurate the results of this procedure ($N = 68.8 \cdot 10^{22}$).

The key to how I am interpreting Perrin rests on my assertion that Perrin's goal in producing values for N is to validate Einstein's equation (E); if that is the case, then his goal is not to argue robustly for the accuracy of his derived values of N using a variety of experimental methods, since it is the methods themselves that are being tested, not the values for N. To further vindicate my interpretation, consider that Perrin expends considerable effort in both Perrin (1910) and Perrin (1916) justifying his assumption that the emulsive grains he is using in his experiments move in a truly irregular fashion. These justifications involve three separate 'verifications' (Perrin 1910, 64–68, and Perrin 1916, 114–119), and with these justifications Perrin feels comfortable applying Maxwell's distribution law to the movement of the grains. Accordingly, he considers himself to be in a position to derive Einstein's formula, once he grants as well the applicability of Stoke's law (which he believes to have been previously shown). The final touch involves experimentally confirming Einstein's equation, which comes about by finding that it produces a value for N 'sensibly equal to the value found for N [in the distribution experiments]' (Perrin 1916, 121; see also the Perrin 1926, 153–154, for similar comments).

Einstein's equation (E) concerns the displacements of Brownian particles. As Perrin notes, there is an analogous equation for the rotations of such particles: Where A^2 symbolizes the mean square of the angle of rotation in time t, and the remaining symbols are as before with (E), we have (Perrin 1910, 73, Perrin 1916, 114, 124)

$$\text{(R)} \quad A^2/t = (R \cdot T) \,/\, (N \cdot 4\,\pi\, a^3\, \xi)$$

As with (E), Perrin's concern is to 'verify' (R) (Perrin 1910, 73, and Perrin 1916, 125), and the method for doing this involves generating values of N, which is possible since all the remaining variables in (R) can be measured. There is a slight complication in doing this, as the rotation is faster given particles of a smaller radius. For instance, with grains 1 micron in diameter, the speed of rotation is 800 degrees per second (Perrin 1916, 125; Perrin 1910, 73, lists a speed of 100 degrees per second, still far too fast for him). A more manageable diameter is 13 microns, but at this size a number of experimental complications appear. In brief, such large-sized grains tend to coagulate, and the only intergranular solution that can alleviate this problem is a urea solution. From here, Perrin reasons as follows. If we begin with 'the probable exact value of N', which he lists as $69 \cdot 10^{22}$ (1916, 126), and if we put in place the conditions we have set forth (involving a urea solution and 13 micron diameter grains), then in applying equation (R) we should expect a value of $\sqrt{A^2} = 14$ degrees per minute. What we find through experimentation is 14.5 degrees per minute, which corresponds to $N = 65 \cdot 10^{22}$. Since this experimentally generated value for N coheres with the expect value of N (as produced through the vertical distribution experiments) within allowable experimental error, it follows for Perrin that Einstein's equation (R) is verified.

Earlier on, we indicated that Einstein in deriving equation (E) made the assumption that the fall of emulsive grains due to gravity can be described by means of Stoke's law. The equation at the basis of this assumption is

$$\text{(D)} \quad D = (\mathrm{R} \cdot T) \,/\, (N \cdot 6\,\pi\, a\, \xi)$$

where D is the coefficient of diffusion (Perrin 1910, 53, 75, and Perrin 1916, 113, 127). Despite having previously justified Stoke's law in his experiments involving vertical distributions of emulsive particles, Perrin wishes to have a more direct confirmation of the law, which he thinks he can do with (D). In Perrin (1916), he examines two cases: the first involving large molecules (in particular, Jacques Bancelin's experiments using sugar solutions) and the second using Léon Brillouin's experimental work on gamboge grains (Perrin 1916, 127–132 ; Perrin 1910, 75–76, looks only at Einstein's work with sugar solutions; Perrin reports that Einstein later revised his work upon hearing of Bancelin's results). Again, the strategy is exactly as we have seen above. As all the variables in (D) can be measured, except for N, we have a way of generating values for N to see whether they cohere with the accepted value (Perrin 1916, 129). Because they do, we establish on firm footing (D) and by extension Stoke's law as well.

TAKING STOCK

We are not yet half way through Perrin's table found at the beginning of this chapter—but we are in a position to foretell the end of the story as regards why Perrin believes he has established an accurate value for Avogadro's number and has demonstrated the molecular view of matter. The bulk of Perrin's work that is original, and that forms the basis for his Nobel Prize, is his work with emulsions and his assumption that there is an informative and useful analogy between the (Brownian) movements of emulsive particles and the Brownian motion of molecules. In this respect, he is carrying through the vision set forth in Einstein (1905):

> In this paper it will be shown that according to the molecular-kinetic theory of heat, bodies of microscopically-visible size suspended in a liquid will perform movements of such magnitude that they can be easily observed in a microscope, on account of the molecular motions of heat. It is possible that the movements to be discussed here are identical with the so-called "Brownian molecular motion"; however, the information available to me regarding the latter is so

> lacking in precision, that I can form no judgment in the matter. (1; quoted in Nye 1972, 112–113)

What Perrin has done is provide this experimental precision. To begin, starting with his vertical distribution experiments, Perrin justifies the claim that there is a useful analogy between uniform emulsions and molecular gases based on his surprising observation that the densities of gases and emulsions each exhibit a geometrical progression as one ascends to greater heights. With this analogy in place, he calculates his best estimate for Avogadro's number (roughly, $68 \cdot 10^{22}$). Armed with this value, he proceeds to find 'verifications' for a number of laws in the 'molecular kinetic theory of Brownian movement' (Perrin 1910, 74): Einstein's equations (E), (R) and (D), Stoke's law and Maxwell's distribution law. So, Einstein continues,

> If the [Brownian] movement discussed here can actually be observed (together with the laws relating to it that one would expect to find), then classical thermodynamics can no longer be looked upon as applicable with precision to bodies even of dimensions distinguishable in a microscope: an exact determination of actual atomic dimensions is then possible. (1905, 2; quoted in Nye 1972, 113)

That is, since the proponent of the classical thermodynamic view—the competitor to the 'discontinuous hypothesis' (Nye 1972, 113)—is not in a position to account for the above (equations and) laws, by justifying these laws one would have provided an effective disproof of classical thermodynamics—which is just what Perrin did. 'On the other hand', Einstein notes, 'had the prediction of this movement proved to be incorrect'—such as if any of the above laws had not been verified—'a weighty argument would be provided against the molecular-kinetic conception of heat' (Einstein 1905, 2, quoted in Nye 1972, 113). We have, then, the reason why Perrin (1916) thinks he has succeeded in establishing the 'real existence of the molecule' through his emulsion experiments (207): These experiments have put on firm footing a body of theory governing the properties of molecules. In other words, molecules are shown to exist 'as described'.

We now continue to work through Perrin's table and examine some of the other approaches Perrin considers for arriving at Avogadro's number, though our discussion here need not be as thorough. As we saw at the beginning of this chapter, Perrin believes that the convergence of the magnitudes 'obtained by each [approach] when the conditions under which it is applied are varied as much as possible' establishes that 'the real existence of the molecule is given a probability bordering on certainty' (Perrin 1916, 207), and we now have an idea why this might be so. The methods Perrin introduces for deriving values of Avogadro's number bring with them assumptions that are part of the molecular theory of matter (they have to, if they are to serve in calculating a value for *N*). In Perrin's experiments regarding the displacement, rotation and diffusion of Brownian particles, these assumptions include (E), (R) and (D), Stoke's law and Maxwell's distribution law, and when the experiments generate values for Avogadro's number that cohere with the values for *N* produced by his vertical distribution experiments, these assumptions are verified. Similarly, the other experiments Perrin adduces involve a wide variety of different sorts of physical phenomena that are also able to generate values for Avogadro's number by means of various molecular theoretic assumptions, and when these values for *N* cohere with the accepted value calculated by Perrin, the molecular assumptions underlying these other sorts of phenomena are 'verified', just as (E), (R) and (D), Stoke's law and Maxwell's distribution law are verified. With each such verification we establish that much more of the body of doctrine comprising the molecular theory of matter. In this way molecular theory is progressively justified and the real existence of molecules 'given a probability bordering on certainty' (Perrin 1916, 207). Let us then examine some of these other physical phenomena that Perrin uses for the purposes of the investigative strategy we just outlined. To start, we see this approach utilized in his discussion of Marian Smoluchowski's molecular theory of critical opalescence. This theory, as it is mathematically formalized by Willem Keesom, generates a prediction for the value of *N*, and, as Perrin (1916) suggests, 'A comparison of the value of *N* derived thus with the value obtained already will therefore enable us to check the theories of Smoluchowski and Keesom' (138). Similarly, Lord Rayleigh's molecular theory explaining the blueness of the daytime sky contains a prediction of the value of *N*, and it is the coherence of this value with Perrin's accepted value as derived from his vertical distribution experiment that leaves Perrin with no doubt that 'Lord Rayleigh's theory is verified' (1916, 142). Again, Planck's quantum-theoretical law of black body radiation contains a prediction for *N*, and Perrin finds a 'striking verification [for this theory lying] in the agreement found between the values already obtained for Avogadro's number and the value that can be deduced from Planck's equation' (1916, 153). However, we need to point out, the investigative strategy we are ascribing to Perrin is not universally applied with all the different kinds of physical phenomena he cites. For example, the language of 'verification' does not occur in Perrin's discussion of Millikan's work on determining the charge on an electron (the 'atom of electricity'). He notes that the value of *N* predicted by Millikan's work is consistent with the value he derives in his emulsion experiments, without suggesting that he is verifying or putting to test Millikan's theoretical assumptions. The same is true with regard to Perrin's discussion of the theory of radioactivity: He is able to generate a number of values of *N* involving different sorts of radioactive phenomena that all agree within experimental error with his preferred value for *N* without claiming that he is 'verifying' or putting to test the theory of radioactivity. There may be a number of reasons for this change in tone. It may be that Perrin is not systematic with his use of the term 'verified'—when he says only that a derived value of *N* is 'consistent with' his accepted value, he may actually mean 'verified', after all. Or perhaps the theories underlying the atom of electricity and radioactivity are so well established that Perrin feels it would be presumptuous on his part to suggest that these theories need further support from a field as distant as colloidal chemistry. Perrin, for his part, does not provide any explanation for his change in terminology where he fails to adopt the language of 'verification'.

Nonetheless, a good proportion of the various physical phenomena he cites have the feature of having their molecular assumptions justified (or 'verified', as Perrin puts it) by generating values for *N* that cohere with Perrin's preferred calculation of *N*. This accordingly gives us an explanation for why these other phenomena are examined—the reason is not to ground a robustness argument for the accuracy of Perrin's initial calculation of *N* that he derived using emulsions. One can find textual support for this interpretation of Perrin's dialectical strategy, a strategy that prioritizes his work with emulsions and that uses this work to test or calibrate other molecular investigations, in the conclusion to his (1910). Perrin says,

> I have given in this Memoir the present state of our knowledge of the Brownian movement and of molecular magnitudes. The personal contributions which I have attempted to bring to this knowledge, both by theory and experiment, will I hope... show that the observation of emulsions gives a solid experimental basis to molecular theory. (92)

It is also an interpretation of Perrin's work that is endorsed by the historians of science Bernadette Bensaude-Vincent and Isabelle Stengershas (1996), who comment:

> To convince the antiatomists, Perrin wanted to find an experimental procedure that was above all suspicion. He found it with the emulsions, by crossing the theory of Brownian motion and van't Hoff's osmotic model. (234)

I now want to argue that a key virtue of reading Perrin this way is that it better explains why he believes his experimental work grounds a realism about molecules.

PERRIN'S REALISM ABOUT MOLECULES

Interestingly, and unexpectedly, Perrin ends both Perrin (1910) and Perrin (1916) by noting the possibility of a nonrealist reading of both his experimental results and the various ancillary observed phenomena he has cited—a reading where, as he says, 'only evident realities enter' (1910, 92; Perrin's italics removed). The result is an instrumentalist approach to molecules, where all reference to molecular reality is removed. To illustrate, recall that Perrin computes *N* using a variety of experimental strategies, each involving characteristic mathematical, functional relationships.

But now, instead of calculating *N*, Perrin suggests we could simply relate the functional relationships themselves while dropping *N*, leaving us with very surprising relationships between, for example, black body radiation and the vertical distribution of emulsive particles. This instrumentalist option is not rebutted by Perrin (1910). On the other hand, Perrin (1916) openly, though somewhat cryptically, rejects such an instrumentalism in the following way:

> We must not, under the pretence of gain of accuracy, make the mistake of employing molecular constants in formulating laws that could not have been obtained without their aid. In so doing we should not be removing the support from a thriving plant that no longer needed it; we should be cutting the roots that nourish it and make it grow. (207)

What Perrin is suggesting, I contend, is that the molecular theory that informs both Perrin's research on emulsions as well as the other sorts of observational phenomena he considers (such as Smoluchowski's molecular theory of critical opalescence, Rayleigh's molecular explanation of the blueness of the daytime sky and so on) cannot be ignored, if we are to understand how these approaches succeed at generating values for Avogadro's number. For instance, Perrin's derivation of *N* based on measurements of the displacement of emulsive particles requires that one can extend the various laws of molecular motion—(E), (R) and (D), Stoke's law and Maxwell's distribution law—to the movements of emulsive particles, and this extension only makes much sense if molecules are thought to be real in the same sense that emulsive particles are real. Moreover, without a realism about molecules, there is no rationale for why Perrin compares his work on emulsions with the work he cites on critical opalescence, black body radiation, the atom of electricity and so on. Finally, absent a realism about molecules, we lack guidance on how one should even interpret *N*.

However, Bas van Fraassen (2009) launches a critique of a Perrin's realist interpretation of molecules, basing his investigation on Perrin (1910) (unfortunately van Fraassen ignores Perrin's *Atoms* since he believes Perrin

1910 is 'much closer to [Perrin's] actual work', a claim he doesn't substantiate—see van Fraassen 2009, 17). Van Fraassen says,

> It is still possible, of course, to also read [Perrin's experimental] results as providing evidence for the reality of molecules. But it is in retrospect rather a strange reading—however, much encouraged by Perrin's own prose and by the commentaries on his work in the scientific and philosophical community. For Perrin's research was entirely in the framework of the classical kinetic theory in which atoms and molecules were mainly represented as hard but elastic spheres of definite diameter, position, and velocity. Moreover, it begins with the conviction on Perrin's part that there is no need at his [sic.] late date to give evidence for the general belief in the particulate character of gases and fluids. On the contrary (as Achinstein saw) Perrin begins his theoretical work in a context where the postulate of atomic structure is taken for granted. (22–23)

Van Fraassen is referring to Peter Achinstein's (2003) book in which Achinstein reads Perrin as using hypothetico-deductive reasoning in support of the existence of molecules. For instance, on the basis of an analogy between emulsive particles and molecules, Perrin derives a value for Avogadro's number by means of his vertical distribution experiments, a value that calibrates the accuracy of other approaches to deriving *N*. For instance, the value for *N* produced by the displacement of emulsive particles is consistent with Perrin's 'preferred' value, a result that accordingly justifies a number of key molecular assumptions, such as Stoke's law and Maxwell's distribution law. With this justification Perrin presumably supports his realist interpretation of molecules. But surely, Achinstein contends, such support is question begging, since the reality of molecules is already assumed with both the vertical distribution and displacement experiments—it is assumed in asserting to begin with that there is an analogy between molecules and emulsive particles. Similar forms of circular reasoning occur with Perrin's examination of Planck's quantum-theoretical law of black body radiation, Smoluchowski's theory of critical opalescence, Rayleigh's theory explaining the blueness of the daytime sky and all the other kinds of physical phenomena Perrin cites. In each case,

Perrin is supporting the reality of molecules by assuming their reality in the context of the theoretical analysis given for each such phenomenon. The problem is to explain how observations, generated under the assumption that there are molecules, can themselves confirm the hypothesis that molecules exist.

We have in fact examined this sort of question in chapter 1 and arrived at the conclusion that nothing prohibits the testing of a hypothesis using observational results that themselves depend on this hypothesis in their generation: In brief, observational results depend in part on the contingent state of the world and so can generate negative results for theoretical hypotheses, even if in generating these results this hypothesis is assumed. But there is another way that Perrin can respond to this problem. Note, to begin with, that each of the listed experimental approaches leading to a calculation of Avogadro's number involve different assumptions applicable to molecules—to take a simple example, the applicability of Stoke's law is key to the Brownian motion experiments but is irrelevant to experiments dealing with critical opalescence. Thus, though each of these experiments assumes the reality of molecules, they assume different things about molecules that may or may not be true. Hence, when Perrin uses the vertical distribution experiments as a standard with which to evaluate the other experiments, what he is doing is testing the correctness of the independent assumptions the other experiments need to make; with the confirmation of these assumptions, Perrin is thus able to build up the molecular theory of matter. From here one might argue (though this isn't necessarily Perrin's argument) that one thereby puts on sound footing a realist interpretation of the molecular theory of matter. Perrin's calibration of the values of N generated from a diverse set of phenomena serves to confirm a variety of different molecular assumptions, with the result that the molecular theory is correspondingly fuller and more detailed. By comparison, a realism about molecules is less justified where there is correspondingly little theoretical development and where what development there is lacks empirical justification—here, one is simply less clear about what molecules are and what the empirical ramifications of their existence amounts to.

But what about the vertical distribution experiments themselves? Can the results of these experiments be said to justify the hypothesis of the molecular nature of matter? The purpose of these experiments, as I have interpreted them, is to generate the best value possible for Avogadro's number. However, for this calculation of N to succeed, there must be a working analogy between a gas and an emulsion, which contingently and fortunately turns out to be the case, since emulsions and gases both exhibit similar exponential distribution laws. On the basis of this analogy we can regard an emulsion as exhibiting (just as a gas does) Brownian motion and from here put ourselves in a position to derive Avogadro's number, since vertical distributions of emulsive particles are observable and thus mathematically representable. So although it is true that the molecular hypothesis is assumed in these experiments, we nevertheless do learn something about molecules—that their vertical distributive properties can be studied by examining uniform emulsions. What this means is that the theory of molecular motion is developed in a way that is empirically testable and as such is a better candidate for a realist interpretation.

Chapter 5

Dark Matter and Dark Energy

In chapter 2 we saw that microbiological experimenters neglected to utilize robustness reasoning, preferring instead to use what we called 'reliable process reasoning'. Ultimately this was for good reason, as the most robust set of experiments, if they lead to any conclusion, would lead to the wrong conclusion—that mesosomes (as bacterial organelles) exist. Similarly, in chapter 3 we saw that WIMP researchers neglected to use robustness reasoning in rebutting DAMA's WIMP-identification claim, despite the fact that arguing in this way would bolster their position. Cases like this constitute evidence that robustness reasoning does not play a substantive role in the justification of scientific observations. Still, the critic of robustness reasoning must face the fact that some scientists, such as Jean Perrin (as we saw in chapter 4), express support for this form of reasoning and that such reasoning has a certain degree of intuitive plausibility. It is accordingly incumbent on the critic of robustness to respond to such concerns.

The answer is revealed by noting the close affinity robustness reasoning has to other, though different forms of reasoning that have an obvious epistemic merit. For example, Perrin views his vertical distribution experiments as providing the most reliable determination of Avogadro's number and uses the results of these experiments to 'verify' other approaches to determining *N*. It follows, then, that the ability of these other experiments to generate values of *N* that converge with the values produced by the vertical distribution experiments shows that they are reliable as well. Such a form of 'calibration' can easily appear as an instance of robustness reasoning. Consider, for example, the value of *N* produced using Smoluchowski's molecular theory of critical opalescence. To answer the question of whether this value is accurate, Perrin shows that his vertical distribution experiments (which act as a standard) retrieve the same value for *N*, and so, by this (perhaps even surprising) convergence, the critical opalescence approach is deemed reliable. Here the style of reasoning being used is without doubt compelling, so long as the assumptions of the case are granted (i.e., the vertical distribution experiments generate an accurate value for *N*, and this value is approximately the same value for *N* as generated using the theory of critical opalescence); moreover, one might be inclined to ascribe the compelling nature of this reasoning to its being a form of robustness. But this would be a mistaken analysis. The situation is similar to the case we cited from Locke in the Introduction, where, in determining whether one is reliably seeing a real fire, Locke counsels us to employ a tactile approach. In this case, if one weren't aware of the details of Locke's reasoning, one might attribute the force of Locke's reasoning to its being a form of robustness reasoning (something Peter Kosso presumably does). But in fact Locke's point is that a tactile approach is just that much better at identifying sources of heat and so a more reliable observational procedure.

To illustrate the matter further, consider the case we examined at the start of the Introduction where we read a newspaper report describing the discovery of alien life. To fill out the case somewhat, imagine there are two local, equally reputable (or perhaps equally disreputable) newspapers that contain the same report on alien life. Would this convergence be a startling coincidence for which we must cite the report's truth as an explanation? Were the report highly contentious, as we can assume it is in this case, it is doubtful that our skepticism would be assuaged much with even convergent reporting once we factor in the usual journalistic standards set by local newspapers—we don't expect news reporters to be experts in (astro)biology or astronomy, and so we anticipate they'll need advice from whomever they deem (fallibilistically) to be experts. Accordingly, our surprise at the coincidence of their reports may ultimately be due simply to our surprise that the two reporters rely on the same purported authority. But however we account for this coincidence, in order to decisively settle the matter (in light of its contentiousness), we eventually need to consult an authoritative source, perhaps the testimony of whichever scientist made the discovery—and even then, because scientists often disagree among themselves about whether a discovery has been made, we will need to examine and evaluate the relevant justification behind the discovery. That is, our strategy should not be to just multiply fallible sources and then explain the surprising coincidence of these sources, but instead to reference an authoritative source that can potentially serve (after suitable scrutiny) as a scientific standard. As such, when we find newspaper reports converging in the way described, and we feel epistemically secure in this reportage, it must be because we think there is a reliable, scientific standard vindicating the accuracy of the report that we implicitly trust. It's doubtful that our epistemic security will be bolstered much by the convergent testimonies of two or more relatively unqualified news reporters.

My goal in this chapter is to look at another way in which scientists can appear to be reasoning robustly, though in fact they are using a different form of reasoning, one that has clear epistemic credentials and in the context of which robustness reasoning can (misleadingly) appear to be epistemically meritorious. This different form of reasoning I call 'targeted testing', and it is similar to robustness in that the empirical justification of a claim profitably utilizes alternate observational routes. How targeted testing differs from robustness, though, is in the strategic nature of the choice of alternate routes: One chooses an alternate route to address a specific observational question that, if empirically answered, can effectively distinguish between two theoretical competitors. In other words, in the absence of this relevant strategic goal, it is not claimed that the reliability of these alternate routes is enhanced should their generated results converge. In what follows I aspire to illustrate the value of targeted testing in two recent, scientific cases. The first case involves a key, empirical proof for the existence of dark matter (i.e., dark matter understood in general terms, not specifically as WIMPs). This proof involves telescopic observations of a unique astronomical phenomenon called the Bullet Cluster that in 2006 largely settled the controversy about whether dark matter exists. The second case deals with the discovery of the accelerative expansion of the universe in the late 1990s (often explained by the postulation of dark energy), for which three individuals—Saul Perlmutter, Brian Schmidt and Adam Riess—jointly received the 2011 Nobel Prize. In this case, the justification for the discovery is based on substantive observations of extremely distant (high redshift) exploding stars, or supernovae. In both the dark matter and the dark energy episode, multiple observational strategies were effectively and decisively utilized—but solely for the goal of targeted testing. Moreover, both episodes contained the potential to exhibit applications of 'pure' robustness reasoning (i.e., robustness unaffiliated with either targeted testing or Perrin-style calibration), yet in neither episode did the participant scientists concertedly argue in this fashion (although in the dark energy episode, one of the lead scientists, Robert Kirshner, made repeated use of robustness reasoning in his popularized account). Overall, these astrophysical episodes are useful to us for the purposes of dimensional balance: Whereas the first three cases dealt with observations of the very small (subcellular structures, subatomic particles and emulsive grains), we now study empirical research into the very large (colliding galaxy clusters and exploding stars).

DARK MATTER AND THE BULLET CLUSTER

In chapter 3 we considered both empirical and theoretical reasons in support of the existence of dark matter. On the empirical side, we noted evidence for dark matter from the rotation curves of spiral galaxies, the velocity distributions of galaxy clusters and evidence from gravitational lensing. On the theoretical side, dark matter is able to help explain large-scale structure formation in the early universe (i.e., the observed temperature and density fluctuations in the cosmic microwave background are so small that, without dark matter, not enough time is available for structure formation; see Nicolson 2007, 47–48); also, dark matter is needed to account for the formation of light elements in the early universe (the so-called Big Bang nucleosynthesis).

Taken as a whole these 'explanatory' justifications for the reality of dark matter have convinced many astrophysicists, despite the presence of some empirical obstacles (e.g., as Nicolson 2007 notes, the existence of dark matter halos implies the existence of 'dark matter cusps' at the center of galaxies, for which we lack empirical confirmation; see 74–76). In addition, the efficacy of these explanatory justifications in generating a belief in dark matter might have persisted had there not been lingering doubts caused by the presence of alternative explanations for, notably, galaxy cluster velocity dispersions and galactic rotation curves. One such alternative explanation proposed by the physicist Mordehai Milgrom involves a change to the theory of gravity as opposed to the postulation of dark matter. Milgrom advocates a theory called MOND (MOdified Newtonian Dynamics; for an introductory review, see Milgrom 2002) according to which an object moving in a gravitational field with sufficiently low acceleration (with a threshold identified by Milgrom) is subject to less gravitational force than an object with a higher acceleration—below this threshold, its velocity will vary with the inverse of the distance rather than the square root of the distance as set forth in the standard Newtonian gravitational force model (Nicolson 2007, 77). What this means is that MOND is able to explain the anomalous rotation curves of spiral galaxies without the invocation of dark matter: Rotation curves simply flatten as we move a large distance from the centre of a galaxy because, at that distance, the force of gravity naturally diminishes. Given also MOND's ability to explain the Tully–Fisher relationship regarding the luminosities of spiral galaxies and its consistency with the observed decrease in the rotation curves of some small elliptical galaxies, it has been claimed by some astrophysicists that MOND is able to stand as a viable alternative to dark matter (see Nicolson 2007, 78, for discussion).

MOND, nevertheless, has its drawbacks. For example, in violating Newton's law of gravity, MOND violates as well the theory of general relativity (Nicolson 2007, 80). In response, a relativistic extension to MOND has been proposed by Jacob Bekenstein called TeVeS (Tensor-Vector-Scalar field theory), which has the added benefit of explaining the gravitational lensing of galaxy clusters without the need to include extra 'dark' mass. TeVeS, moreover, can account for large-scale structure formation without invoking dark matter (see Dodelson and Liguori 2006), something beyond the capacity of MOND. Considering as well that the hypothesis of dark matter is itself not beyond empirical reproach (mentioned above), there has been a perceived need in the astrophysical community to definitively decide between MOND (and other modified gravity approaches) and the dark matter hypothesis.

From the perspective of a philosopher of science, the MOND/dark matter controversy is an interesting test case for how scientists resolve problems of theoretical underdetermination. There is no suggestion here, of course, that MOND and the dark matter hypothesis are empirically equivalent: Though each are empirically compatible with (and can indeed explain) the observed rotation curves of spiral galaxies, there is evidence against dark matter that is not evidence against MOND (e.g., the absence of dark matter cusps) and evidence against MOND that is not evidence against dark matter (e.g., the detectable mass of some 'dark galaxies', galaxies containing only hydrogen gas and no suns—see Nicolson 2007, 79). Furthermore, in the astrophysical community there is a decided bias in favour of the dark matter hypothesis—MOND is definitely the underdog hypothesis as evidenced by the fact that worldwide there are numerous research ventures directed at detecting dark matter particles, such as the WIMP detection experiments we discussed earlier, but a negligible number of experiments directed at detecting changes in the force of gravity at low accelerations. Nevertheless, MOND has posed enough of a challenge for astrophysicists to attempt to resolve the dark matter/MOND controversy once and for all.

A breakthrough in this regard occurred in 2006 via a group of astrophysicists led by Douglas Clowe. In a publication describing their work, Clowe, Randall, et al. (2006) note the existence of 'alternative gravity theories', such as MOND, that can be used to 'reproduce at least the gross properties of many extragalactic and cosmological observations' (1), such as the observed rotation curves of spiral galaxies. Prior to 2006, this dialectical situation had left the astrophysical community in somewhat of a stalemate: Scientists, Clowe and colleagues claim, were left 'comparing how well the various theories do at explaining the fine details of the observations' (1), that is, looking for minute differences in observational data that could effectively distinguish between competing theories (such as predicting with greater precision a galaxy's rotation curve). Clowe, Randall, et al. never expressly state what is misleading about such an approach. We can conjecture that, if the debate is to be fought over the fine details of observations, then each theory will always have the option of adjusting its parameters so as to accommodate these details—and a definite refutation of one of the approaches will never be had. Neither do they see the point of a robustness approach. For example, in 2005 a colleague of mine in my university's Physics and Engineering Physics Department described to me the sort of robustness argument one could use as an evidential basis for dark matter (Rainer Dick, personal correspondence). He writes:

> The evidence for dark matter seems very robust. It arises from different methods used by many different groups: galaxy rotation curves,

> gravitational lensing from galaxies and galaxy clusters, observations of the peculiar velocities of galaxies and galaxy clusters, magnitude-redshift relations for type 1a supernovae, peaks in the angular correlation of anisotropies of the cosmic background radiation.

However, despite such 'robustness', the astrophysical community at that time was not fully convinced about the existence of dark matter in the face of alternatives such as MOND. In the end it was only the unique evidence provide by Clowe, Randall et al. that settled the matter. This evidence established the existence of dark matter, something a robustness argument could not do.

So what was Clowe, Randall, et al.'s (2006) special evidence for the existence of dark matter? What they sought to do was locate a situation in which dark matter is 'physically separated from visible matter' and thus detectable 'directly by its gravitational potential' (1; see also Clowe, Bradac, et al. 2006, L109). Throughout the universe, dark matter omnipresently pervades visible matter; galaxies and everything else in them float in vast dark matter 'halos'. Since the signature for the presence of dark matter is its gravitational potential, showing the existence of dark matter usually involves an inference based on an observed discrepancy between the amount of normal, luminous matter one sees in a galaxy and the amount of matter one infers to be present from witnessing a galaxy's gravitational field, say, by looking at a galaxy's rotation curve (if we are examining an elliptical galaxy). Of course, the need to perform such an inference underpins the underdetermination problem that faces the choice between the hypothesis of dark matter and the hypothesis of modified gravity, for the observed features of a gravitational field (such as the relevant rotation curve) can be explained both by invoking dark matter and by assuming an alteration in the force of gravity. This sets the stage for Clowe, Randall, et al's (2006) ingenious solution to this problem—they propose to resolve this evidential stalemate by identifying an astrophysical situation in which, by fortunate happenstance, dark matter is physically (and not just conceptually) separate from luminous matter.

To this end they utilize a unique astrophysical phenomenon, called the Bullet Cluster, whereby two galaxy clusters (each containing potentially many thousands of galaxies) have collided in the plane of the sky and are at the point where they have just passed through one another. Images of the Bullet Cluster taken by Clowe, Randall, et al. (2006) are the product of two sorts of telescopic methods. First, optical images (generated from the Hubble Space Telescope) record the visible light emanating from the galaxies that constitute each galaxy cluster. Light is also recorded from the stars and galaxies forming the cosmic backdrop to the cluster; this light is useful because, as it passes by the Bullet Cluster, it is bent by the gravitational field produced by the cluster with the result that the shapes of these stars and galaxies are distorted to some degree. This phenomenon is called gravitational lensing, and it is by measuring the extent of these distortions of the shape of background stars and galaxies that one can reconstruct and map the gravitational field of a lensing cosmological object, such as the Bullet Cluster. With lensing we can produce a contour map with 'higher altitudes' denoting a stronger gravitational potential (and thus a more massively dense source), with surrounding plateaus indicating drop-offs in such potential. Now, with a galaxy cluster like the Bullet Cluster, the majority of the gravitational potential where we are considering only *luminous* matter rests not with the galaxies themselves but with a hot x-ray-emitting gas that pervades a galaxy cluster, called the intra-cluster medium (ICM). This medium cannot be detected by a light telescope, such as the Hubble, so the Chandra X-ray Observatory is used to track the ICM. In the resultant, computer-generated image combining both optical and x-ray data, one sees three areas of color. First, we can see the white light of two groups of galaxies comprising the galaxy clusters that have just passed through one another (galaxies are said to be 'collisionless'; they do not interact with one another when the clusters to which they belong collide). Second, blue light in the generated image represents areas of maximum gravitational potential reconstructed from the gravitationally lensed, distorted images of the stars and galaxies that form the backdrop of the Bullet Cluster. Here we find two such areas of blue light spatially coinciding with each of the two sets of visible galaxies. By contrast, these areas of coincident white and blue light are clearly separated from two pink areas signifying the locations of intense x-ray emissions, representing the ICMs for each of the colliding galaxy clusters. These pink areas trail the galaxies because, unlike the galaxies themselves, they collide (i.e., they aren't collisionless)—so much so that the ICM of one of the colliding clusters forms a '(pink) shock front', giving it the appearance of a bullet (hence, Bullet Cluster). We now have a surprising, unexpected result: The bulk of the mass of the Bullet Cluster does not reside where the bulk of the luminous mass resides (i.e., the ICMs); rather, it resides in a location coincident with the galaxies themselves and as such is not accounted for by these galaxies, since the galaxies form a very small part of the gravitational potential of a galaxy cluster. At this point Clowe, Randall, et al. (2006) and Clowe, Bradac, et al., (2006) state that they have found what they call 'direct evidence' for the existence of dark matter, evidence that conclusively repudiates the modified gravity approach.

What do Clowe and colleagues mean when they say they have 'direct evidence' for the existence of dark matter? The evidence from the Bullet Cluster phenomenon, they say, is direct in the sense that it '[enables] a direct detection of dark matter, independent of assumptions regarding the nature of the gravitational force' (Clowe, Bradac, et al. 2006, L109; see also Clowe et al. 2004, 596, and Clowe, Randall, et al. 2006, 1). Recall that, with the Bullet Cluster, the areas of greatest gravitational potential—the areas where the mass of the cluster is most concentrated—are spatially offset from the areas where the luminous mass is concentrated—the ICMs for each of the colliding clusters. Accordingly, one can modify the gravitational force law as MOND demands but not change the fact that the bulk of the mass for each of the clusters that make up the Bullet Cluster is at a different location than the respective ICMs of these clusters, which is assumed to make up the majority of the luminous mass of a galaxy cluster. Thus, even if we permit the possibility of an alternative gravitational theory, this does not remove the support the Bullet Cluster provides for dark matter. Even granting the truth of such an alternative theory does not change the fact that the bulk of the mass of the cluster does not lie with the bulk of luminous mass.

In what way is this evidence for dark matter better than the explanatory justifications described earlier? Consider, for example, the justification for dark matter on the basis of the high rotational velocity of the outer edges of spiral galaxies. MOND and TeVeS both count this phenomenon in their favour because they are able to theoretically account for it, and so if a dark matter theorist wishes to use this phenomenon to justify the existence of dark matter, these alternative theories need to be discounted beforehand—which leaves the theorist in no better a position than before. Things, though, are different with the Bullet Cluster: Here it doesn't matter if one assumes one of the alternative gravity theories, or lacks a reason to discount them beforehand, for we have evidence on behalf of the existence of dark matter independent of the status of these alternative theories. By comparison, a modified gravity theory has few options to circumvent the empirical fact that, with the Bullet Cluster, the source of gravitational potential does not correspond to the location of the majority of luminous mass (few, but not zero, since a modified gravity theory could potentially account for the apparent displacement of mass, given sufficient, albeit unorthodox conceptual flexibility). As a consequence, the Bullet Cluster evidence has been successful in convincing even those in the modified gravity camp about the reality of dark matter. The originator of MOND himself, Moti Milgrom (2008), comments:

> We have known for some fifteen years now that MOND does not fully explain away the mass discrepancy in galaxy clusters.... Even after correcting with MOND you still need in the cluster some yet undetected matter in roughly the same amount as that of the visible matter. Call it dark matter if you wish, but we think it is simply some standard matter in some form that has not been detected.

Of course, neither MOND nor any of the other alternative gravity theory excludes *necessarily* the existence of dark matter (i.e., in the above quote Milgrom sees MOND as embracing the existence of dark matter). In fact, Clowe and colleagues do not claim to irrevocably disprove a modified gravity theory by introducing the Bullet Cluster evidence. Instead, the question is whether there is compelling evidence to believe in the existence of dark matter—evidence that holds even assuming the truth of a modified gravity theory—and the Bullet Cluster is purported to provide 'direct' evidence in this regard.

The line of reasoning Clowe, Bradac et al. (2006) and Clowe, Randall et al. (2006) advocate is an example of what I call targeted testing. It is similar to robustness in that the empirical justification of a claim utilizes an alternate observational route, yet the choice of alternate route is strategic: It has the specific goal of addressing an observational question that, if empirically answered, can effectively distinguish between two competing theoretical hypotheses. With the Bullet Cluster evidence, for example, we have observational proof that dark matter is distinct from luminous matter, yet the value of this evidence does not rest in the fact that it is another, independent line of justification, for there are already a variety of different lines of justification one can use to this effect. Rather, the value of this evidence is that it provides a proof for dark matter that holds even given the truth of a key theoretical competitor, the modified gravity hypothesis. In the absence of this strategic advantage, the Bullet Cluster evidence wouldn't have settled the dark matter issue for astrophysicists just as, historically, finding convergent, independent evidence did not succeed in doing this.

It is not hard to find targeted testing being used analogously in a number of the episodes we examined in previous chapters. We saw it used earlier in the mesosome case where microbiologists target tested assumptions that underlay their experimental methods. Consider, for example, the two competing hypothesis that mesosomes are natural features of bacterial cells versus the possibility that they are unnatural, derivative features of damaged, sickly bacteria. The fact that bacteria when placed in unusual, even toxic environments exhibit mesosomes—such as when exposed to anesthetics, antibiotics and anti-microbialpolypeptides—target tests this pair of alternatives and speaks against the reality of mesosomes in normal cells. Note, on the other hand, that, from a robustness perspective, this evidence supports the existence of mesosomes, as we have significantly different observational procedures jointly and independently exhibiting the presence of mesosomes. Similarly, with DAMA's WIMP detectors, a key question is whether DAMA is witnessing an annual modulation of WIMP detection events or, alternatively, only a modulation in the local amounts of radon gas. DAMA suggests a way to target test this alternative, which involves tracking the modulation in the local concentration of radon to see if it mimics the observed modulation in detection events, and by this means they are able to counter such a possibility by observing that the modulation of ambient radon gas does not synchronize with their observed modulation results. Note again the irrelevance of robustness reasoning here. If we found an annual modulation of radon gas to mimic DAMA's observed result—which would indeed be a surprising convergence that involves independent observational procedures—this would of course not speak on behalf of DAMA's pro-WIMP claim.

My plan now is to give a further example of the evidential significance of targeted testing drawing from a related area of astrophysical research, the investigation into the accelerative expansion of the universe that leads to the postulation of the existence of dark energy.

This historical episode is valuable for two reasons. First, it is a case in which one would anticipate the use of robustness reasoning by scientists, since the discovery of the accelerative expansion of the universe involves the work of two research groups arriving independently at the same observational results. In fact, some of the key participants in these groups describe themselves as reasoning in a robust sort of way. Nevertheless, as I show, when one looks in detail at how the scientists in this case reason, they do not reason robustly after all (just as Perrin wasn't arguing robustly, when one looks in detail at his reasoning). Rather, the main justificatory support for the universe's accelerative expansion involves the use of targeted testing. The second value of this dark energy case is simply its unmistakable quality as state-of-the-art scientific research, given the fact that the lead members of the two research groups very recently received Nobel Prizes for their discovery (in December 2011). Using the history of science to illuminate a philosophical point can run the risk of using outdated or marginal science; this is pointedly not the case with the discovery of the accelerative expansion of the universe leading to the postulation of dark energy.

TYPE IA SUPERNOVAE AND DARK ENERGY

To begin our discussion of recent research into the accelerative expansion of the universe, it is worthwhile recounting some of the historical background to this research. The first major breakthrough occurred in the 1920s when Edwin Hubble found evidence that the universe was expanding. A critical element to Hubble's work was his use of cepheid variables as cosmological distance indicators (i.e., as 'standard candles'). Cepheid variables are stars that pulsate at different periods depending on their brightness; brighter stars pulsate with longer periods. Thus when one observes the sky at night and sees two cepheid variables pulsating with the same frequency, one knows that the fainter star is farther away and that it isn't, instead, just an intrinsically fainter star. With his knowledge of cepheid variables, Hubble could estimate the distance of galaxies by identifying cepheid variables in these galaxies. Another important aspect of Hubble's investigation was his determination of the redshift of galaxies. It is possible to recognize when, and to what degree, light emanating from a galaxy is shifted to the red. The explanation for this phenomenon is that the wavelength of light is stretched by the movement of the galaxy away from us (the viewers), just as sound waves are stretched and exhibit a lower pitch when an object emitting a sound travels away from us (i.e., more stretching, and so a redder color or lower pitch, corresponds to a faster recession velocity). What Hubble did was to relate these two variables: the distance of a galaxy and its recession velocity. To this end he graphed a relation, called a Hubble diagram, which shows clearly that a galaxy's redshift increases with the distance of the galaxy—the farther away the galaxy, the faster it is receding from us. From this diagram it became clear that the universe is expanding. (For background on Hubble's work, see the introductory discussions in Nicolson 2007, 21–23, and Kirshner 2004, 67–70.)

Although cepheids are bright, they are not bright enough to serve as useful distance indicators for the distances cosmologists need to investigate in order to determine the expansion history of the universe. (As Kirshner [2004] notes, we need to examine the redshifts of galaxies '1 or 2 billion light-years away', whereas cepheids are only useful up to 50 million light-years; 103). Enter a new and different distance indicator, Type Ia supernovae (SN Ia), which are exploding stars 100,000 times brighter than a cepheid (Kirshner 2004, 104; there are other types of supernovae, including II as well as Ib and Ic, which are not used as standard candles; for an informative, accessible review, see Nicolson 2007, 116–117). The source of the value of SN Ia rests not just in their tremendous intrinsic brightness but also in the fact that such explosions generate light that follows a characteristic pattern: First, the light follows a typical brightness curve, taking about 20 days to arrive at a peak intensity and then approximately 2 to 3 months for the light to subside; second, the exact dimensions of this curve depend on its peak brightness—a brighter SN Ia will have a light curve with a more prolonged decline. SN Ia are thus similar to cepheids in that we can ascertain their brightnesses on the basis of a feature that is easily and directly measurable: for cepheids, their brightness is indicated by their period; for SN Ia, brightness is determined using the shape of their light curves.

Since the 1980s, SN Ia have been increasingly used to extend the Hubble diagram to higher redshifts and larger distances from us in order to measure the universe's expansion rate at times further in the past. (In an expanding universe, objects at higher redshifts are further away from us, and so in examining them we are looking further into the past because of the time it takes for the light of these distant cosmological objects to reach us. Hence, redshift can be used as a measure of time—an object viewed at a higher redshift is an object that existed at an earlier stage of the universe). The first research group to make effective headway in this task was the Supernova Cosmology Project (SCP), formed in 1988 under the leadership of Saul Perlmutter. This headway was matched by a second group, the High-Z Team (HZT; *z* stands for 'redshift'), organized in 1994 by Brian Schmidt and Nick Suntzeff. (See Kirshner 2004 for a useful and candid recounting of the history of the work of these two teams; Filippenko 2001 is similarly valuable, written by someone who had associations with both teams.) It is the competing work of these two groups that eventually formed the basis of the discovery of the accelerative expansion of the universe in 1998 and thence to the postulation of dark energy as the purported cause of this expansion. Dark energy is in fact a generic term for whatever it is that causes the accelerative expansion of the universe. A common view is that dark energy is the 'cosmological constant', a mathematical artifice invented by Einstein in 1917 to reconcile general relativity theory with the assumption (current at the time) that the universe was static, that is, neither expanding nor contracting (see Kirshner 2004, 57–58). Einstein envisaged the cosmological constant as providing an 'expansive tendency to space' (Kirshner 2004, 58), one that was no longer needed once it became accepted (following Hubble) that the universe was expanding. But it now seems to many astrophysicists that Einstein's artifice needs to be resurrected in order to accommodate (once more) the 'expansive tendency of space'. Unfortunately, such an interpretation of dark energy has proved problematic since Einstein's cosmological constant, strictly speaking, entails an expansive tendency to space of the order of 10^{120} too large, given what is needed to accommodate the observed accelerative expansion of space (see Caldwell and Kamionkowski 2009, 589, and Perlmutter 2003, 2470, for more on this problem with the cosmological constant in accounting for the expansion of space).

Let us now look more closely at the work of SCP and HZT that led to their discoveries that the expansion of space is accelerating and that therefore dark energy exists. The efforts of both groups involve examining SN Ia at high redshifts and measuring both the intrinsic brightness of SN Ia (by examining their light curves) as well as their apparent brightness (discerned by using a light telescope, such as the Hubble Space Telescope). To orient their work, they consider various models for the expansion of the universe. One particular model takes precedence, which we call the 'received model' (due to its adoption by a majority of astrophysicists), according to which the mass density of the universe is not so great as to halt the universe's expansion but that gradually this expansion will decelerate until it stops in the infinite limit. This density (whatever it turns out to be) is called the 'critical' density and is given the arbitrary value 1 with the symbolization $\Omega_m = 1$. A different model is one in which the mass density of the universe is less than 1 ($\Omega_m < 1$). On this model, the expansion of the universe is decelerating but not quite as fast as with the received model, and thus in this universe the expansion does not stop, not even in the infinite limit. Finally there is a 'coasting' universe, which is void of any matter ($\Omega_m = 0$); a coasting universe maintains its expansion unretarded since there is no counter-effect due to the gravitational force of matter.

Given these various models of an expanding universe, SCP and HZT proceed as follows. Suppose we have located an SN Ia with a redshift of a certain value. Now let us take the two extreme cases: $\Omega_m = 1$ (a 'flat' universe; whereas with $\Omega_m < 1$ we have an 'open' universe) and $\Omega_m = 0$ (a coasting universe). The redshift of this SN Ia indicates that it is moving away from us at a particular velocity, and whereas in a coasting universe it has always moved away from us at that velocity, in a flat universe, because the universe's expansion is decelerating, the universe has not expanded as much as it would have in a coasting (or open) universe. This means that the SN Ia would be brighter in a flat universe as compared to a coasting (or open) universe, as the light from the SN Ia had a shorter distance to travel in order to get to us. From this point both SCP and HZT have the tools to arrive at estimates of Ω_m from observations of the brightness of various SN Ia. Given an SN Ia at a particular redshift, and given the assumption that Ω_m has a particular value, we arrive at an estimate of how bright this SN Ia should appear to be (i.e., brighter if Ω_m is larger). We then observe how bright this SN Ia really does appear to be and from here test our assumption about the value of Ω_m .

With this procedure in mind, SCP collected data during the mid-1990s on a number of Sn Ia and in 1997 published results on seven of them with modestly high redshifts ($z > .35$). They arrived at a value of $\Omega_m = .88$ (we omit error ranges for simplicity) assuming 'a $\Lambda = 0$ cosmology' (Perlmutter et al. 1997, 565, 579). Λ signifies the cosmological constant or, more generically, dark energy. At this stage of cosmological theorizing (in 1997), no one believed in the existence of dark energy. Still, it was recognized that if space had an 'expansive tendency', this would necessarily affect the value assigned to Ω_m. In addition to considering a $\Lambda = 0$ cosmology (i.e., $\Omega_\Lambda = 0$), Perlmutter et al. (1997) also consider the case where Λ had a non-zero value, and, with their data, if $\Omega_\Lambda = .06$ then $\Omega_m = .94$. Either way, their results confirmed the received view at the time that the universe was flat with Ω_m near 1.

Perlmutter et al.'s (1997) conclusions about Ω_m were soon disputed by the HZT group. In Garnavich et al. (1998), four SN Ia were considered: three near $z = .5$ and a fourth with a significantly higher value for z (= .97). Using this data, Garnavich et al. concluded that in a $\Lambda = 0$ universe $\Omega_m = -.1$, which is clearly a physical impossibility. Conversely, if Λ had a non-zero value, then $\Omega_m = .3$ (or .4, depending on what process is used to analyze light-curve shapes). That is, they considered their data to be 'inconsistent with a high matter density' universe, one where Ω_m is near 1 (Garnavich et al. 1998, L56). This was, at the time, a completely novel and unexpected result. SCP, for their part, once they had data for a significantly high redshift SN Ia ($z = .83$), revised (in Perlmutter et al. 1998) their initial view about a high matter density universe and suggested that, for a $\Lambda = 0$ universe, $\Omega_m = .2$ (and $\Omega_m = .6$ if Λ is non-zero). HZT regarded these new results by SCP to be 'marginally consistent' with their data (Filippenko and Riess 1998, 38, and Riess et al. 1998, 1033), but of course there was a key difference in that, for SCP, in a $\Lambda = 0$ universe Ω_m was still greater than zero, whereas for HZT it was a completely unphysical, negative number. Subsequent work by SCP, presented at a pivotal meeting of the American Astronomical Association in January 1998, brought their results in line with HZT's—with results from 40 SN Ia, SCP yielded $\Omega_m = -.4$ under the assumption that $\Omega_\Lambda = 0$. At the same time, HZT revised their estimations to $\Omega_m = -.35$ if $\Omega_\Lambda = 0$, and $\Omega_m = .24$ if $\Omega_\Lambda \neq 0$ (and assuming as well that the universe was flat).

The next question was how to interpret these results, and here I will suggest that there is first of all a simple interpretation and alternatively a more complex one. The simple interpretation is as follows. What the data tell us is that if the universe is flat, then there must be some extra material in the universe apart from matter (both luminous and dark). It is this sort of interpretation that was bandied about in late 1997: As reported by Glanz (1997), many astrophysicists at that time were prone to accept that there must be some form of 'extra material' making up a significant fraction of the density of the universe to make up the gap left if $.2 < \Omega_m < .4$. In reflecting on what this extra material could be, it was standardly assumed to be Einstein's cosmological constant (i.e., dark energy, symbolized by Λ). No other candidate was ever suggested. To this end, the argument for dark energy became almost a straightforward question of addition: $\Omega_m + \Omega_\Lambda = 1$, so if $\Omega_m = .3$, then $\Omega_\Lambda = .7$ (i.e., dark energy exists). To buttress this argument, the following additional lines of argument could be added. First of all, why must the total density be 1? Why must the universe be flat? In support of this conclusion, both SCP and HZT adduced observations of the angular fluctuations of the Cosmic Microwave Background (CMB) by COBE (COsmic Background Explorer) in the early 1990s and subsequently by WMAP (Wilkinson Microwave Anisotropy Probe) launched in the early 2000s, both of which supported the flatness claim (see Perlmutter 2003, 2470, Kirshner 2004, 250–251, 264–265, and Riess et al. 2004, 665; for background review see Nicolson 2007, 107–113). Also, should we expect Ω_m to have a value of .3? Here, SCP and HZT referred to measurements of the mass density of galaxy clusters that confirmed this value (see Perlmutter et al. 1999, 583, Riess 2000, 1287, Perlmutter 2003, 2470, and Kirshner 2004, 264). We have as a consequence a suggestive three-pronged convergence of results: The SN Ia observations lead us to assert the existence of dark energy if the universe is flat and $\Omega_m = .3$; the COBE and WMAP observations confirm the flatness hypothesis; and finally the galaxy cluster observations support $\Omega_m = .3$. As a result, we have a strong argument for dark energy.

This convergence of results left a strong impression on a number of the participant astrophysicists. Saul Perlmutter (2003), for example, describes it as a 'remarkable concordance' (2470); Robert Kirshner (2004), in reflecting on this convergence, notes: 'When completely independent paths lead to the same place, it makes you think something good is happening' (264); such 'agreement [has] the ring of truth' (265; see also 251). It looks like astrophysicists are being convinced about the reality of dark matter by means of a form of robustness reasoning.

In fact there is potentially another form of robustness reasoning one could provide here, one that makes reference to the (eventual) convergence of the results generated by SCP and HZT. For instance, Perlmutter et al. (1999) comments: 'To [a] first order, the Reiss et al. [i.e., HZT] result provides an important independent cross-check for [our conclusions regarding dark energy]... since it was based on a separate high-redshift supernova search and analysis chain' (583). In addition, on behalf of HZT, Filippenko (2001) remarks:

> From an essentially independent set of 42 high-*z* [SN] Ia (only 2 objects in common), the SCP later published their almost identical conclusions (Perlmutter et al. 1999).... This agreement suggests that neither team had made a large, simple blunder! If the result was wrong, the reason had to be subtle. (1446)

Nicolson (2007) presents a very straightforward expression of this robustness view:

> The close agreement between the results obtained by two independent groups, based on largely independent sets of supernova... was truly remarkable and compelled the scientific community to treat the evidence very seriously. (122)

In the end, however, Nicolson is somewhat equivocal about the efficacy of such robustness reasoning. He notes that in 2003 SCP generated data on 11 SN Ia using a process that was intrinsically more reliable than the process that generated the previous data; regarding the former process, Nicolson (2007) remarks that

> it allowed [SCP] to calculate the extent to which supernovae had been dimmed by the obscuring effects of clouds of dust (dust extinction) within host galaxies [with the result that this data was] on its own…good enough to confirm—independently of all previous results—the acceleration of the universe and the need for dark energy. (123)

So Nicolson's view seems to be that, where we have an intrinsically more reliable observational process, considerations of robustness become less significant—indeed, both forms of robustness reasoning to which we have here referred, that is,

1. the independent convergence of empirical data regarding the flatness of the universe, the measurement of $\Omega_m = .3$ using galaxy clusters and the SN Ia observations, and
2. the independent convergence of the SCP and HZT SN Ia observations,

never really convinced the astrophysical community that it should embrace the reality of dark energy. That had to wait until certain forms of systematic error (discussed below) could be effectively controlled.

This leads us to the second, more complex interpretation of the results described above in which SCP and HZT found SN Ia data leading to the conclusion that we live in a $\Lambda \neq 0$ cosmology. It is one thing to say that we live in a low mass universe and that in order to subsidize the cosmic density to ensure that we live in a flat universe we need to include a new form of substance (called dark energy, the cosmological constant or what have you). This is what we conclude from the simple interpretation of the SN Ia results. It is another thing to say the substance making up this lack is a form of repulsive gravity that actually counteracts the gravitational force of mass. On the simple interpretation all we could conclude is that the expansion of the universe is decelerating more slowly than if $\Omega_m > .3$; by comparison, on the second interpretation, if the repulsive gravity generated by this new substance is sufficiently powerful, we could conclude that the expansion is decelerating more slowly than expected on the first interpretation, or that it is even accelerating. Accordingly, if we could observationally confirm a decreasing deceleration, or better still an acceleration of the universe's expansion, this would provide us with more definite proof that dark energy exists, *qua* repulsive gravity, and that it makes up the apparent gap in density in the universe.

This second interpretation of the observed result, that we live in a low mass-density universe, accordingly requires a more precise determination of the expansion rate of the universe to determine if it differs greatly from what we expect if $\Omega_m = .3$. As the pivotal research paper on the topic (Riess et al. 1998) describing observations of 34 SN Ia at wide range of redshifts reveals, it not only turns out that

> the distances of the high-redshift SNe Ia are, on average, 10%–15% farther than expected in a low mass density ($\Omega_m = .2$) universe without a cosmological constant,

(an even more profound result than if we assume $\Omega_m = .3$), but that

> high-redshift SNe Ia are observed to be dimmer than expected in an empty universe (i.e., $\Omega_m = 0$) with no cosmological constant. (1027; italics removed)

In other words, the expansion rate is comparable to what we would expect if the universe contained only a sort of negative mass that had an accelerative effect. This result is echoed in Perlmutter et al. (1999) on the basis of 42 SN Ia of varying redshifts, even though their conclusion is less forcefully put:

> The data are strongly inconsistent with a $\Lambda = 0$ cosmology, the simplest inflationary universe model. An open, $\Lambda = 0$ cosmology also does not fit the data well. (565)

Here, SCP is additionally careful to explain away its 1997 result supporting a high density universe, a result it writes off as due to the influence of a statistically anomalous SN Ia. Omitting this SN Ia (and thus leaving a sample of only 6 SN Ia), Perlmutter et al. (1999) assert that the 1997 data actually cohere with their new data within one standard deviation (582–583). This sort of ad hoc, revisionary assessment of past data is not necessarily an illegitimate maneuver for scientists to make, if the noted SN Ia really is anomalous.

It is on the basis of this second interpretation of the low mass-density result, and the correlative determination that the observed mass density does not adequately account for the expansion rate of the universe, that astrophysicists were convinced to take the dark energy hypothesis seriously. But there were some crucial obstacles to both SCP and HZT resting content with the conclusion that dark energy exists. Even though they had compiled, altogether, a fairly large sample size of SN Ia, thus minimizing the potential for statistical error, there was nevertheless the pressing problem of possible systematic errors (see Riess et al. 1998, 1009, where this point is made explicitly). In the next section we examine such systematic errors and scrutinize how SCP and HZT proposed to handle them.

DEFEATING SYSTEMATIC ERRORS: THE SMOKING GUN

In essence, the SN Ia data collected by SCP and HZT led researchers to the conclusion that dark energy exists because it reveals the SN Ia to be dimmer (less luminous) than expected and not only in a low mass-density universe but in a no mass-density universe as well. The explanation for this dimness is that the SN Ia are farther away than anticipated, which would be the case if the universe's expansion were accelerating. This leads us to the conclusion that the universe contains a source of repulsive gravity, or dark energy, counteracting the attractive gravitational force of matter that retards the expansion of the universe.

But could the extra dimness of the SN Ia be due to another cause? Perhaps there is some systematically misleading factor that is giving the illusion of accelerative expansion? Both SCP and HZT spend substantive time in their research papers considering such possible systematic effects that could mimic dimness. Two key possible sources of error are:

1. Evolution: SN Ia at higher redshifts are older, and perhaps as time progresses the properties of SN Ia change ('evolve'). For example, the chemical compositions of the stars that end up as SN Ia ('progenitor stars') might be different due to differences in the abundances of elements in the universe at that time, and this difference might lead to intrinsically dimmer SN Ia (see Kirshner 2004, 225–227, and Nicolson 2007, 123).
2. Extinction: By extinction, astrophysicists mean the presence of microscopic, interstellar particles, or 'dust', that affect the light we see coming from cosmic objects (see Kirshner 2004, 227–230, and Nicolson 2007, 124). Note that there is both red dust and grey dust to be considered, the former particles being smaller and having a characteristic tendency to 'redden' light and the latter having no reddening effect—it simply dims.

There are in fact a number of other systematic effects to consider, such as the Malmquist bias and other selection biases, K-corrections and gravitational lensing—but SCP and HZT believe that evolution and extinction are the key sources of error that need to be addressed.

SCP, in generating its high mass-density result as described in Perlmutter et al. (1997, 578), as well as its low mass-density result recounted in Perlmutter et al. (1998, 53), asserts that extinction does not have a major influence on its results and so it declines to correct for it. For instance, SCP contends that

> correcting for any neglected extinction for the high-redshift supernovae would tend to brighten our estimated supernova effective magnitudes and hence move [our results] ... toward even higher Ω_m and lower Ω_Λ than the current results. (Perlmutter et al. 1997, 578)

In other words, on SCP's view, a high mass result would be confirmed even further if corrections were made for dust. HZT, by contrast, is critical of SCP for not correcting for extinction: HZT comments,

> Not correcting for extinction in the nearby and distant samples could affect the cosmological results in either direction since we do not know the sign of the difference of the mean extinction. (Filippenko and Riess 1998, 39; see also Riess et al. 1998, 1033)

HZT is similarly wary of the effects of evolution and much more cautious than either Perlmutter et al. (1997) or Perlmutter et al. (1998):

> Until we know more about the stellar ancestors of [SN] Ia, we need to be vigilant for changes in the properties of the supernovae at significant look-back times. Our distance measurements could be particularly sensitive to changes in the colors of [SN] Ia for a given light curve shape. Although our current observations reveal no indication of evolution of [SN] Ia at $z \approx 0.5$, evolution remains a serious concern that can only be eased and perhaps understood by future studies. (Riess et al. 1998, 1033)

By comparison, SCP is less concerned about the prospect of evolution. As regards

> both the low-redshift and high-redshift supernovae... discovered in a variety of host galaxy types,... [the] small dispersion in intrinsic magnitude across this range, particularly after the width-luminosity correction, is itself an indication that any evolution is not changing the relationship between the light-curve width/shape and its absolute brightness.... So far, the spectral features studied match the low-redshift supernova spectra for the appropriate day on the light curve (in the supernova rest frame), showing no evidence for evolution. (Perlmutter et al. 1997, 579)

SCP's apparent laxity on the matter of evolution comes through in Perlmutter et al. (1998) by means of its suggestion that, by examining a singular SN Ia at $z = .83$,

> [high red-shift SN Ia] can be compared spectroscopically with nearby supernovae to determine supernova ages and luminosities and check for indication of supernova evolution. (53)

But determining the effects of evolution (and extinction) is unlikely to be so straightforward. SCP seems to concede this point in Perlmutter et al. (1999):

> Some carefully constructed smooth distribution of large-grain-sized gray dust that evolves similarly for elliptical and spiral galaxies could evade our current tests. Also, the full data set of well-studied [SN] Ia is still relatively small, particularly at low redshifts, and we would like to see a more extensive study of [SN] Ia in many different host-galaxy environments before we consider all plausible loopholes (including those listed in Table 4B) to be closed, (582)

where Table 4B (with the heading 'Proposed/Theoretical Sources of Systematic Uncertainties') lists 'evolving gray dust', 'clumpy gray dust', 'SN Ia evolution effects' and 'shifting distribution of progenitor mass, metallicity, [and] C/O ratio' (582) as potential sources of systematic error.

One reason I have entered on this digression concerning the impact of systematic errors on the evidence for accelerative expansion of the universe is to highlight, in its rough outline, the style of reasoning in which both SCP and HZT are engaged. It can be said that both are involved in what in chapter 2 I called 'reliable process reasoning', though here of a negative sort: So long as the systemic effects of extinction and evolution can be controlled for, telescopic observations of the dimness of SN Ia form a reliable basis on which to assert that the expansion rate of the universe is accelerating; however, since these systematic effects aren't adequately controlled for (given what SCP and HZT knew at the time), it follows that the telescopic observations of the dimness of SN Ia don't form a reliable basis on which to assert the accelerative expansion of the universe. The fact that fields as disparate as experimental microbiology and telescopic astrophysics converge so centrally in the rationales they use at justifying (or dismissing) observed results is perhaps surprising. But perhaps this is not surprising, considering the obviousness and generality of the rationale—in essence the 'reliable process' rationale is simply, 'one should use a reliable observational procedure in concluding that a procedure is generating a true result'. Again, a reliable process rationale is not meant to denote a particularly extraordinary form of reasoning: Simply, a scientist identifies a process as reliable (or not) in terms of producing true reports with inputs of a certain kind and then reports that one actually has an input of this kind, leading to the conclusion that the report is truthful (or not). As was noted earlier, it is left as an open variable what to regard as a reliable process, but that is only because we leave it to the scientists themselves in the context of their respective fields to fill out these details. As it happens, with SCP and HZT, the relevant reliable process is one that corrects for the effects of evolution and dust—and conceivably this could be a process that exhibits robustness. But robustness isn't used by these groups at this stage, just as it is seldom if ever used in the other historical episodes we have studied in this book.

It is worthwhile noting that neither group in fact cites any particular piece of empirical evidence that supports the view that such evolution and dust effects even occur. Rather, such sources of error are simply hypothetical possibilities that need to be excluded if the observed, extra dimness of SN Ia is to ground an argument for the accelerative expansion of the universe and from there the existence of dark energy. Along these lines, consider the appraisal of these problems expressed by HZT member Adam Riess (2000):

> The primary sources of reasonable doubt are evolution and extinction.... Although...[one] could potentially yield evidence that either of these noncosmological contaminants is significant, the current absence of such evidence does not suffice as definitive evidence of their absence. Our current inability to identify the progenitors of [SN] Ia and to formulate a self-consistent model of their explosions exacerbates such doubts. Even optimists would acknowledge that

> neither of these theoretical challenges is likely to be met in the near future. (1297)

In a sense, then, the situation is analogous to the underdetermination problem facing supporters of dark matter versus a theory of modified gravity. We have two hypotheses that can be used to capture the extant evidence between which we cannot rationally choose—either the accelerative expansion hypothesis or the evolution/dust systematic error hypothesis. The goal then, as with the Bullet Cluster case, is to target test these theoretical alternatives. This sets the stage for the subsequent, pivotal telescopic investigations made by SCP and HZT that do, in fact, rule out the problems of extinction and evolution. The resultant decisive evidence is called by some astrophysical researchers the 'smoking gun' (e.g., Filippenko 2001, 1447, and Kirshner 2004, 234).The theoretical basis of the smoking gun is the following insight (see Riess 2000, 1297, Filippenko 2001, 1447, Perlmutter 2003, 2471, Riess et al. 2004, 666, and Nicolson 2007, 124–128, for discussion). The expanding universe immediately following the Big Bang is extremely dense with matter, so dense that the expansion would decelerate even in the presence of dark energy. However, as time goes on and as the mass density attenuates with the continuing expansion of the universe, the dark energy eventually becomes enough of a factor to reverse this deceleration, leading to the accelerative expanding universe in which we currently live. Thus, while looking at SN Ia that are far away (at high redshifts), we should notice the extra dimness of such SN Ia since the universe's expansion is accelerating. However, at one point, especially far from us, we should notice that the SN Ia are instead *brighter* than they would be in an accelerating universe; these would be SN Ia that we observe to exist during the time when the universe's expansion was decelerating. The observational task then is to examine these high redshifts SN Ia to determine their relative brightness. This task was accomplished by HZT in the early 2000s, and the results published in Riess et al. (2004). It was then confirmed that these distant SN Ia were brighter at a redshift of about .5 and higher. The value of .5 signifies the distance to us (or, alternatively, the elapsed time) from the point at which the expansion of the universe moved from decelerating to accelerating, a shift called a '(cosmic) jerk'. The key to this confirmation is that such a brightening would be highly improbable if the dimness of SN Ia that occurred after this cosmic jerk is ascribed to either interstellar dust or SN Ia evolution. Let us assume that the influence of dust or evolution is monotonic—that is, if dimming occurs due to either source, then the farther away the SN Ia, the greater the effect of the dust or evolution, and so the greater the dimming. With dust, the monotonicity of the effect is easy to conceptualize—the greater distance, the more intervening dust, the more dimming. With evolution, too, it is somewhat improbable that the changes progenitor stars (for SN Ia) underwent from the time of the jerk that lead to intrinsically dimmer SN Ia would have gone the other way prior to the jerk and lead to intrinsically brighter SN Ia. The point in either case is that it becomes substantially more difficult to account for dimmer-than-expected SN Ia using the effects of dust extinction and evolution if we are faced with brighter-than-expected SN Ia found to exist prior to the cosmic jerk. As Riess et al. (2004) express the point:

> The data reject at high confidence simple, monotonic models of astrophysical dimming that are tuned to mimic the evidence for acceleration at $z \approx 0.5$. These models include either a universe filled with gray dust at high redshift or luminosity evolution $\propto z$. More complex parameterizations of astrophysical dimming that peak at $z \approx 0.5$ and dissipate at $z > 1$ remain consistent with the SN data (but appear unattractive on other grounds) (686),

an unattractiveness that Riess (2000) calls a 'conspiracy of fine-tuning' (1297).From here it should be clear that the effectiveness of the smoking gun in demonstrating the reality of dark energy is analogous to the way in which the Bullet Cluster demonstrates the reality of dark matter. Given that the dimness of the SN Ia (after the jerk) can be accounted for using either the dark energy hypothesis or either the extinction or evolution hypotheses, the strategy of targeted testing seeks to find an observed result that would support the dark energy hypothesis, even if one were to assume the occurrence (and monotonicity) of extinction and evolution effects. This is what the observed, extra brightness of ancient, pre-jerk SN Ia can provide us. We can, if we like, assume that extinction and evolution effects are in play in our observations of these SN Ia—but this would only mean that these SN Ia are *even brighter* than anticipated to be, since all extinction and evolution do is dim the SN Ia. So, just as with the Bullet Cluster and dark matter, with the extra brightness of pre-jerk SN Ia and dark energy we have found an effective observational strategy for resolving a key underdetermination problem—we have found a way to empirically discriminate between the option that the observed results are due to an accelerative expanding universe (and correlatively dark energy) versus the option that the results are due to some systematic effect.

ROBUSTNESS IN THE DARK ENERGY CASE

A few themes, familiar from our previous case studies, arise from our discussion of the discovery of the accelerative expansion of the universe (and the related discovery of dark energy). The main theme is that, ultimately, robustness reasoning (to the extent that it occurs at all) is not fundamental to the thought processes of discoverers. In this regard, the reader may be unconvinced, given that Robert Kirshner, Saul Perlmutter and others were candidly impressed by the fact that the SCP and HZT groups independently arrived at similar observed results (i.e., that SN Ia are dimmer than expected). The independence of the methodologies used by the two groups is not insignificant. As Kirshner (2004) remarks,

> The distant supernovae [examined] were, with a few exceptions…, not the same. The data reductions were done by different methods. The ways that light-curve shapes were employed to correct for the variation in SN Ia brightness were different. We handled dust absorption in different ways. (222)

One would think that, with such a striking convergence of results, an effective argument for dark energy could have been made strictly on the basis of this convergence. But that is not what happened. In the key research articles to the discovery, one doesn't find this (or any other) robustness reasoning introduced in any authoritative fashion: The convergence of results is stated as more of an afterthought, introduced after the 'real' work of adequately justifying one's observational methods is accomplished.

Most especially, as we noted above, this convergence did not succeed in settling the issue of whether there are sources of systematic error that need to be addressed. It was only after the discovery of results discerning the extra brightness of pre-jerk SN Ia that the authenticity of the accelerative expansion of the universe (and the reality of dark energy) was firmly established; moreover, this discovery stemmed mainly from the work of HZT (e.g., as described in Reiss et al. 2004), not on the basis of results convergently arrived at by both HZT and SCP.

It is also true that, despite the occasional pronouncements of some of the participant scientists indicating the relevance of robustness to their thinking, the details of the historical course of events casts doubt on the efficacy of robustness reasoning in this episode. As Kirshner himself describes the matter, there was significant jostling with the two groups regarding who would make the 'pronouncement' of the accelerative expansion of the universe first. As it happens, the HZT group pronounced first (on 27 February 1998), with much reflective consternation (see Kirshner 2004, 221). It was only later that SCP jumped on board, delayed apparently by its distress over whether it had adequately accounted for the problem of cosmic dust. So the initial justificational basis to HZT's pronouncement was not a robustness argument after all, since one group (SCP) had not, at that time, even committed itself to the result. In the published literature, SCP's advocacy of accelerative expansion occurred a year after the key HZT paper (Perlmutter et al. 1999, as compared to Reiss et al. 1998), and, as we remarked earlier, the relevant SCP paper contains a somewhat ad hoc revision to (contrary) results presented previously in Perlmutter et al. (1997). Thus, there is room here to question even the independence of SCP's result, insofar as it seems to be following HZT's lead. Still, at least one senior member of the SCP team refuses to see SCP as trailing HZT in establishing the accelerative expansion of the universe. Gerson Goldhaber in discussing HZT's work in April 1998 comments, 'Basically, they have confirmed our results. They only had 14 supernovas and we had 40. But they won the first point in the publicity game' (quoted in Kirshner 2004, 221).

Apparently Goldhaber sees SCP and HZT as engaged in a sort of competition, quite the opposite from viewing them as reasoning robustly on the basis of mutually supportive, observed results.

As I argued above, the better way to understand the reasoning that forms the basis to the observed accelerative expansion of the universe is to view it as a form of targeted testing: When we are faced with competing theoretical interpretations of extra-dim SN Ia (again, their dimness is explicable either by their extended distance or by the effects of evolution or extinction), the observations made by Reiss et al. (2004) (i.e., of the extra-brightness of pre-jerk SN Ia) settle the matter by finding evidence that supports the presence of accelerative expansion, even if we assume the occurrence of evolution and extinction. That is, Reiss et al.'s (2004) results target test the possibility that evolution or extinction are the cause of the SN Ia data. A further, even more general description of Reiss et al.'s methodology is to describe it as a form of reliable process reasoning, where the reliability of the observational methods used in determining the extended distance of SN Ia is assured by discounting the impact of various systematic errors such as evolution and extinction. Yet, however one describes the observational strategies of astrophysicists in this episode, it is nevertheless clear that the form of reasoning that ultimately decides the issue of the universe's accelerative expansion (and the attendant argument for dark energy) is not a form of robustness reasoning, a fact unaltered even if we regard the convergence of HZT's and SCP's observed results as surprising.

Chapter 6

Final Considerations Against Robustness

Our case studies—the mesosome, the WIMP, Perrin's atoms, dark matter and dark energy—reveal that robustness lacks the methodological pride of place many philosophers (and many scientists in their reflective moments) attach to it. Scientists often ignore robustness arguments when they have obvious application (such as in the WIMP episode where various research groups employing model-dependent approaches could have but failed to use robustness reasoning); sometimes they describe themselves as using robustness arguments when in fact they are doing something else (such as with Jean Perrin's arguments for the existence of atoms). Overall I hope to have shown that robustness reasoning does not play much of a role in how scientists justify their observed results. My task now is to further my philosophical critique of robustness, inspired in part by the historical case studies we have been examining. In what follows I provide a variety of considerations leading to the cumulative conclusion that there is very little, if any, value to be found in 'pure' robustness reasoning, reasoning that considers it an epistemic merit to multiply independent observational procedures leading to an observed result, even though this multiplication serves no additional purpose (e.g., in order to 'target test' as in the dark matter and dark energy cases or to 'calibrate' as in the Perrin case). To begin, I return to a consideration of the core argument formulated in chapter 1, an argument that, as we saw, forms the basis to many probabilistic attempts to justify the value of robustness reasoning.

INDEPENDENCE AND THE CORE ARGUMENT

The core argument for robustness states: If independent, observational processes converge on the same observed result, this puts us in a position to cite both the representational accuracy of this result and the reliability of the processes as a way of explaining this convergence. As we elaborated this argument, if an observational report is the product of two (or more) different physical process (or, in epistemic terms, the product of two or more distinct theoretical assumptions), then there is less of a chance the report is only an artifact of one of these processes (or simply a byproduct of one of these assumptions) since the independent production of the same artifact, despite a change in the physical process (or in the assumptions used), is highly unlikely. In such a case, we would tend not to suppose that one or other of the processes (or one or other of the assumptions) is uniquely responsible for the production of this report (i.e., that the report is the result of some physical or theoretical 'bias'). Instead, it is assumed, there must be some other explanation for this produced report, presumably the reliability of the processes that generate this report along with this report's truth.

Clearly, the crux to this argument is the assumption that the physical processes under consideration are independent (henceforth we leave aside for simplicity epistemic forms of independence, as the arguments will apply to them, *mutatis mutandis*). Although there is no denying that physical processes could be independent, we are nevertheless left with the problem of determining when, in fact, processes are independent in a way that is suitable to ground a robustness argument. Steve Woolgar (1988) expresses the difficulty as follows (here by 'triangulation' he means 'robustness'):

> The essence of triangulation ... is that knowledge arises from *different* representations of the *same* thing.... However, ... 'sameness' or 'difference' is not an inherent property of (sets of) phenomena. (80; Woolgar's italics)

Let us put Woolgar's point this way: Our judgment that we have found different observational procedures that converge on the same observed report is a theoretically significant one, for the sameness or difference of these procedures is not obvious from bare inspection. For instance, take the case where I utter the observational report, 'This is fire', at 10:00 AM. Also suppose that, because I am uncertain about whether I am really seeing a fire, I check to see whether I am prompted to utter the report, 'This is fire', at 10:01 AM, and then at 10:02 AM and so on. All else being equal, these subsequent checks doubtfully add much epistemic weight to my claim 'This is fire', for few would consider checking at 10:00 AM, at 10:01 AM, at 10:02 AM and so on to be different, independent procedures. But how do we know this? That is, if we are queried, 'Why are these routes the same?', can we say that we simply observe this sameness? I don't think it would be that easy. One could just as well observe the difference in these procedures by pointing at the different times at which they occur, noting the subtle change in the weather patterns at each subsequent minute and remarking on the slightly different orientations of the physical components of the procedure relative to the sun and moon. Don't these differences make for different and independent observational procedures, and so don't they provide the grounds on which to base a robustness argument?

Here, in defending the nontriviality of robustness, one might suggest that the cited differences aren't relevant—that the issue of what time it is, what the weather is like and our astronomical orientations are irrelevant to determining whether a fire is present. But of course this need not be true. For example, it may be that someone is subject to periodic hallucinations of fire but that these hallucinations seldom last, and so if the appearance of fire remains after one or two minutes, one can be sure it wasn't hallucinatory. Or suppose it starts to rain heavily at 10:01 AM and the fire, despite being exposed to the weather, isn't extinguished; then this change in weather really does matter to our assessment that there was a (real) fire there one minute ago. The point is that whether two (or more) observational procedures are the same or different, and, if they are different, whether they are different in a way that matters for the purpose of the proper evaluation of an observed report, is not a straightforward matter and would require in every case a certain degree of theoretical or empirical acumen.

How then might we go about assessing the relevance of alternative observational procedures? It probably goes without saying that any relevant observational procedure, alternative or not, will need to meet some sort of reliability standard: No observational procedure will be relevant if it's patently unreliable. But if we have dispensed with any probabilistic notion of independence, as we have suggested we must do in chapter 1, then there's not much else to guide us from a robustness point of view as regards the independence of observational procedures. Hacking (1983) says robustness involves 'completely different physical processes' (201). But they mustn't be too different, such as using a thermometer to tell the time or a clock to measure the temperature (and if there were a convergence of reports in such cases, it would doubtfully tell us anything informative, despite the startling nature of the convergence). Perhaps we should say that different processes must at least be about the same subject matter, in the sense that seeing a fire and feeling a fire are both about fire, whereas clocks are about time and thermometers are about temperature. In this sense, seeing a fire at 10:00 AM and seeing it again at 10:01 AM are both about fire; moreover, both processes have assured reliability and are (at least numerically) different physical processes, so perhaps we have here a working case of robustness after all. But robustness theorists would likely dismiss the value of such reasoning as regards seeing a fire at 10:00 AM and then seeing it again at 10:01 AM, as they would not consider these processes different enough. So for worthwhile robustness reasoning, there's presumably a need for alternative observational procedures that are different enough, yet not too different—and here there's no guidance at all on how this medium amount of difference is to be determined. More important, it's hard to see what a medium amount of difference has to do with an assessment of relevance. It might be that the closer in details one observational procedure is to another, the more relevant their respective results are to each other, say in a case where the goal is replication. Alternatively, it might be that the results of one observational procedure are highly relevant to the results of another procedure precisely because the procedures are so different, as might be the case when one calibrates an observational procedure with another observational procedure that, as it happens, is much different (such as when one calibrates an electron microscope with a light microscope, taking the latter as authoritative where the levels of magnification overlap).

As opposed to analyzing when observational procedures can be said to be different in the right degree to be both independent and relevant, productive methods for assessing the significance of alternate observational procedures were revealed in our case studies, namely, though calibration and targeted testing. The former involves using a procedure whose reliability is assured as a way of confirming ('verifying', in Perrin's parlance) the results of other procedures, a practice that can enrich one's theoretical understanding of a common subject matter of these procedures. The latter identifies a weakness in the informativeness of standard observational processes, a weakness that leads to an uncertainty in the theoretical significance of the results (e.g., one cannot rationally decide between two empirically adequate, though conflicting, theoretical competitors) and in response institutes a new, alternative observational procedure that effectively and decisively target tests this weakness and so clarifies the theoretical situation. But these forms of reasoning take us far beyond the presumed insight that is the basis for the core argument for robustness. With the core argument, when independent observational processes converge on the same observational result, this apparently puts us in a position to infer the representational accuracy of this result and the reliability of the adduced processes as a way of explaining this convergence. Again, the underlying idea is that if an observational report is produced by means of two different physical processes, then we can't attribute this result to some bias in one or other of these processes that individually produces this report. But the notion of independent though still relevant alternative observational procedures lacks clarity, both when we interpret this notion probabilistically (as we saw in chapter 1) and nonprobabilistically (as we see here). Moreover, both calibration and targeted testing—the reasoning strategies we suggest can effectively address the relevance issue—are arguably ways of approaching observational reliability that entrench, and do not avoid, theoretical biases: In cases of calibration, the reliability of one observational procedure is upheld as a standard for other procedures, and in targeted testing we adopt a preference for one observational process due to its unique ability to distinguish theoretical alternatives. In both of these types of cases, it isn't a convergence of results that establishes the joint reliability of two (or more) procedures (along with the accuracy of their observed results) but rather the established quality of one procedure that can calibrate/target test other procedures.

On the basis of these sorts of reasons, I believe that the core argument is ultimately unsuccessful. I now want to deepen my critique of robustness by addressing some lingering, relevant issues. We start by considering further the value of independent observational sources.

THE NEED FOR INDEPENDENCE DOES NOT EQUAL THE NEED FOR ROBUSTNESS

As we saw in the dark energy case, both the SCP and HZT teams were occupied with measuring the faintness of high redshift SN 1a with the goal of testing their models of an expanding universe. Kirshner (2004) applauds this fact, commenting, 'All along we had made the case that it was a good thing for two independent groups to carry through this work' (222). However, Kirshner doesn't explain why having two (or more) independent groups is a good thing. Presumably his view is that the value of independent work rests on the possibility of generating robust results, and he does in fact claim that because the results generated by SCP and HZT converge such results have the 'ring of truth'. So let us consider a situation where one research group, in reflecting on the work of another research group who is investigating the very same topic, speculates on whether it should adopt a similar physical procedure as the other group or a different physical procedure. At first glance, there is no particular value in using a different physical process or different assumptions 'just for the sake of it'. At the very least, whichever process is used, it has to meet a minimal reliability condition and the adduced assumptions presumably have to be both true and relevant. But more than that, if the physical (observational) process one has adopted is believed to be the most reliable of all the various observational processes that could be considered, because perhaps the theoretical assumptions that underlie this observational process have the greatest likelihood of truth or are the most relevant, then why would one want to utilize, alternatively, processes or sets of assumptions that are any less than this? If a research group reflects on the work of another group using different physical processes and decides that the quality of this other work does not match the quality of their own work, why would the first group even bother itself with the work of the others? For instance, in the dark energy case, HZT didn't regard its own work as having adequately demonstrated the universe's accelerative expansion until various systematic errors were properly handled—so why would it look for assurance to the work of SCP when SCP hadn't even itself accounted for these errors? *Pace* Kirshner, it's not clear why it's a good thing for two (or more) independent groups 'to carry through this work'.

Yet let us step back a bit and reflect on why, in generating an observational result, a research group would decide to carry out an investigation that is independent of the work of other groups. In the first place, what does it mean to carry out independent work? One suggestion is that, when we have two research groups (A and B), A's work is independent of B's work if A is not aware of what B is doing (and vice versa), or perhaps A is aware of what B is doing but ignores this information, shutting it out of A's (collective) mind. That would explain their respective states of surprise when they arrive at the same results; something else must be driving the convergence of their results than their (perhaps unconscious) mutual awareness. However, one imagines that maintaining such a state of independence in real scientific practice would be quite difficult. Members of research groups working on the same topic often meet at conferences, have liberal access to each other's publications (say, by acting as peer reviewers for publications and grants) and even on occasion switch from one group to another (as Alex Filippenko did, going from SCP to HZT). Thus it is hard to think that researchers could effectively remain independent in this way—each group would soon find out if a competing group was close to achieving a key result, could easily learn about what methods the other group was using to generate the result and might find itself highly motivated to achieve the same result as a matter of priority. Of course one might suggest that being aware of another group's work is one thing and letting that group's work affect one's own work is another. But it may be difficult to establish that one is not being so influenced: One may need to delve into the subconscious minds of researchers to determine if they have been unconsciously influenced, even if they openly disavow such an influence. Even if one could perform this psychological inquiry, one may wonder whether for the purposes of assessing the accuracy of an observed result this is a worthwhile activity. With the need to ascertain the independence of observational methods, one would expect scientists who were proponents of robustness to recruit the services of psychologists to confirm the independence of a researcher's thinking from her possible awareness of a competitor's work. It hardly needs to be said, though, that such psychological inquiries seldom occur in the sciences (the exceptional case is when there's the possibility of fraud) and that generally scientists would look askance at the perceived need to perform such a psychological investigation. For them, determining that an observation report is reliably generated depends not on whether the user of the procedure is aware of others using this procedure but on whether the procedure that produced the report is of high quality. A scientist will ask, Is an observational procedure theoretically well grounded, well calibrated, error-free and so on? It won't matter to them (other than for moral reasons) that others use this procedure and that this has perhaps influenced them to use the procedure as well. After all, how does not being influenced in this way make the procedure more reliable? Intuitively, independence in the sense of not being influenced by the work of others is not significant at all in establishing the reliability of an observational procedure, and in fact by being aware of how others use this procedure one could learn how to work out some of the procedure's 'bugs' or perhaps gain insight on how one should 'tweak' its protocols.

Still I think we can say that, in a case where research groups are unaware of what each other is doing, there is a motivational benefit to be had in aspiring to such independence in that each group is impelled to rely on its own resources to complete the observational task at hand. There is, in a sense, a prohibition on a sort of cheating—one can't cheat by finding out a competitor's (important) results and then ensuring that one's own results are in sync with them. Similarly, there is a prohibition on studying a competitor's apparatus and then copying his method, pretending that one has arrived at this method on one's own. Rather, each group must determine independently how to make the relevant observations and must base its decision regarding the worth of its observational method on the inherent reliability of this method as determined by a slate of factors, such as the ability to remove sources of systematic error, ensure the sensitivity of instruments, maintain model independence (as in the DAMA case), justify (perhaps empirically) the theoretical assumptions underlying an observational strategy (as in the mesosome case) and so on. To be sure, as we noted, it can be enormously difficult to remain ignorant of a competitor's work, so one would expect there to be a residual influence on one's own work. Nevertheless, and ideally, there is a benefit to cognitive independence (if we can call it that) in terms of the challenge it presents to researchers to resolve observational issue on their own and be innovative in their thinking—for it is in being so challenged that novel and productive ideas are often generated. Here Perrin's work is a case in point. His work with emulsions was quite unique, based as it was upon an observational strategy he developed independently of other researchers. It is ultimately because of this uniqueness that he was awarded the Nobel Prize—not for his having reproduced the reliable results already generated by others. Indeed, we can find in all our episodes a similar recognition on the importance of independent thinking in this sense: Mesosome researchers went beyond the standard R–K methodology to methods that employed freezing, DAMA ventured out with a unique model-independent strategy, Clowe et al. focused on an entirely new astronomical phenomenon and HZT sought data at extremely high redshifts never before witnessed. As these cases illustrate, thinking independently has enormous value for empirical scientific research.

Now I believe there is a sense of 'independent work' where one could say that independent work has a definite informational advantage: It is a case where separate inquiries generate separate pieces of information that, put together, allow one to draw an inference unattainable from each piece of information taken by itself. A good example of this advantage is found in the dark energy case. In that case, we noted the independent convergence of empirical data regarding (a) the flatness of the universe (using CMB measurements), (b) measurements (using galaxy clusters) of Ω_m that give a value of .3, and (c) SN Ia observations that reveal the expansive acceleration of the universe. Gates (2009) describes the situation this way:

> As the twentieth century came to a close, [the] situation changed dramatically. Three independent kinds of observations of the Universe (with several groups working independently on each kind of observation) now provide compelling evidence for a flat Universe whose major component is some form of dark energy. (198)

Here the pieces of information are independent in the sense that they concern different subject matters: Measurements of the CMB are different from cluster mass measurements, which are different again from measurements of the faintness of distant SN 1a. Altogether these pieces of information lead one to infer the existence of dark energy (though not irrevocably as we noted above, since there are other ways to explain the faintness of distant SN 1a than by assuming the presence of dark energy). However, this is not an example of robustness reasoning, even though it is an example of using independent sources of information. This is because the independent sources are generating distinct, separate pieces of information, whereas the characteristic feature of robustness is that the same piece of information is generated using different (convergent) methods. It would not be, for example, an example of robust reasoning to conclude that Socrates is mortal by inferring this claim from the independent assertions that Socrates is a man and that all men are mortal. Similarly it is not robustness reasoning to use a device to observe some entity and to then adduce additional empirical considerations to confirm the good working order of this device. For example, we saw Silva et al. (1976), Dubochet et al. (1983) and Hobot et al. (1985) all using empirical considerations to justify their approaches to fixing biological specimens, just as the various WIMP research groups used empirical checks to ensure the accuracy of their WIMP detectors. But in neither of these cases do we have a form of robustness reasoning, because in both cases we have an observational procedure investigating some (possible) phenomenon (such as mesosomes or WIMPs) and then an additional observational procedure whose subject matter is something entirely different, to wit, the original observational procedure. By contrast, when Kirshner (2004) says that 'it was a good thing for two independent groups to carry through this work' (222), he does not mean the work of empirically and reflexively testing one's observational procedure (which does have an epistemic value). Rather, he means using different physical procedures (or adopting different theoretical assumptions) to perform the same observational task (such as measuring the faintness of high redshift SN 1a)—the trademark of reasoning robustly. So even in those cases where independent sources of evidence are found to be epistemically valuable, they turn out not to be cases that fit the style of robustness reasoning.

THE CONVERSE TO ROBUSTNESS IS NORMALLY RESISTED

Now one would think that, if robustness were a valuable indicator of the reliability of an observational process, conversely the failure of robustness should be a valuable indicator of the nonreliability of such a process. In other words, one would think that, where an observational result fails to be robust—that is, where an alternative, at least minimally reliable observational process generates a contrary result to the original observational process—then this should signal to us the possibility that the result is not reliably generated and that indeed we should regard the result as false, or at least unjustified. Call this 'converse robustness'.

It turns out that in our case studies we can find some resistance among scientists to reasoning in this way. For example, in the WIMP detection case, DAMA's annual modulation result was not confirmed by any of the alternative model-dependent approaches, but despite that divergence, DAMA wasn't inclined to discard its results—rather, it critiqued the informativeness of these other approaches. The model-dependent groups acted in the same way: Though their results differed from DAMA's, that didn't lead them to question the quality of their own experiments. Instead, they raised challenges for DAMA's modulation strategy. Of course such behavior is entirely reasonable if in each case the alternative approaches lack authority. Yet we are dealing here with high-level research groups in the relevant area of investigation, groups that are well published and well funded. Consider similarly that in the mesosome case, once frozen-hydration and freeze-substitution approaches to preparing bacterial specimens began to be used as alternatives to the standard R–K approach, and such specimens were found not to display mesosomes, microbiologists did not immediately repudiate the existence of mesosomes as they should by converse robustness. Instead, their subsequent study turned to identifying the most reliable approach in investigating bacterial substructure, with some groups persisting with the R–K method (and its exhibition of mesosomes), and some other groups adopting newer approaches and either ignoring the testimony of the R–K approach or arguing that it contains flaws (at least as regards the investigation of bacterial substructure). Finally, a surprising case of resistance to converse robustness, a case where robustness fails but this failure doesn't redound to the unreliability of the original process, occurs in Kirshner (2004). Kirshner, who otherwise overtly supports the value of robustness reasoning, comments:

> We worried a little that the LBL team [i.e., SCP] had published a contrary result [to ours, i.e., HZT's]. But this was hard work, and there were many ways to go wrong. We decided not to worry too much about the other guys, to judge our own measurements by our own internal standards, and to hope for the best. (208)

These comments remind us of DAMA's rationalization of the many ways in which model-dependent approaches to detecting WIMPs can go wrong; but whereas DAMA has no stake in robustness reasoning, Kirshner apparently does. His rationale for dismissing contrary, observed results sounds to me disingenuous: One would think that if he were a dedicated proponent of robustness, and if robustness, whenever it occurs, has (as he suggests) 'the ring of truth', then the fact that competing strategies (each meeting the minimum of reliability) generate different results should, for him, speak contrariwise against the reliability of his own results. Yet Kirshner provides no specific reasoning for dismissing these results, saying only in the above quote that with the generation of such results there are 'many ways to go wrong'. One is reminded of the UKDM group in the WIMP case that, as we noted in chapter 3, disavows the benefit of retrieving the same results as other WIMP-detection groups on the grounds that considering the work of other groups only increases the 'uncertainty' in the data. In Kirshner's hands, this consideration seems to work to insulate his group's work from refutation by the results of other groups (in contrast to UKDM, who could have alluded to the similar no-WIMP detection results generated by other model-dependent groups).

To illustrate what is at stake here, consider by comparison cases of replication that are superficially like cases of robustness, though ultimately different from robustness in a crucial respect. The demand for replicability can be expressed as follows: If an observed result is retrieved by means of an observational process that is asserted to be reliable, then this result should be derivable by other scientists using the same process in different circumstances. (It is acknowledged that sometimes replicability is not feasible because of the uniqueness of the circumstances that generated the result; consider, for example, the nonreplicability of the observation of the return of Halley's Comet in 1758, as predicted by Newtonian mechanics.) If these other scientists fail at replicating the result, then this highlights a need to scrutinize the observational procedure for its reliability. For instance, researchers might investigate the circumstances in the first case under which the observed result was generated to determine whether these circumstances are adequately reconstructed in the replicated case. If the replicated conditions are then more adequately reconstructed and the observed result still doesn't appear, it is incumbent on the researchers to determine whether there are certain unforeseen circumstances in the second case that might be thwarting a successful repetition or circumstances in the first case that are artificially producing an observed result. The key point for us is that it wouldn't make much sense to simply disregard a failure of replication, claiming that this is 'hard work' in which there are 'many ways to go wrong', to not 'worry too much about the other guys' and simply 'judge our own measurements by our own internal standards'. Such reasoning doesn't play with replication—and it shouldn't play with robustness.

Let me note, nevertheless, that there is a crucial difference between replication and robustness. What is being sought in replication is a new observational procedure that mimics as closely as possible the original one. In this respect, it would be ideal if the original procedure could be repeated identically, but because of the necessary limitations on exactly repeating an observational procedure it follows that the circumstances of the replication will of necessity vary somewhat from the original run of the procedure (e.g., a replicated experiment at the very least will occur at a different time). As such, the inherent variations in replicated data can be viewed as unfortunate byproducts of statistically variable observational procedures. By comparison, with robustness what are sought are different observational procedures that don't just mimic the original one but that involve fundamentally different physical processes. Variations in the generated data could therefore be the result of these systemic differences and not just a result of statistical variance. Still, one might think of replication as in fact involving an application of robustness reasoning since the replicated circumstances necessarily vary somewhat from the original circumstances, say by occurring at a different time, in a different place or with different scientists. But the difference between replicated results and results that are robust is made clear when we consider a case where the same result is successfully attained under replication. Here the conclusion that the replicated result is correct is based on the fact that the original process is reliable, along with the claim that the result really does issue from this process, as shown by the fact that the result comes about when the process is repeated as exactly as possible. By comparison, with robustness, the conclusion that the observed result is correct is based on the belief that both the original process and a novel process generate the same result. These differences, however, don't mask the fact that what is being deployed in both cases are observational procedures that the proponents believe are reliable (only one procedure with replication, two or more with robustness). Accordingly, when contrary results are generated, such as when a replication fails or when varied observational processes fail to generate the same result, astute observers have the epistemic duty to diagnose these failures and not simply dismiss them, whether these observers are engaged in replicating a result or reasoning robustly. It is then because scientists, as I have suggested, are somewhat dismissive of converse robustness (though not dismissive of contrary replications) that I am left with the impression that they are not really active proponents of robustness reasoning—despite occasionally speaking on behalf of robustness, as Kirshner does.

THE CORROBORATING WITNESS: NOT A CASE OF ROBUSTNESS

There's been a crime, and the police officer is interviewing potential witnesses who can identify the perpetrator. Witness 1 describes the individual as a short, stocky man with a thick black mustache. Is the witness reliable? The police officer looks around for an independent witness (e.g., one who isn't simply mimicking the first witness) and locates Witness 2 who, like the first witness, asserts that the perpetrator is indeed a short, stocky man with a thick black mustache. The police officer now feels confident that the first witness is reliable and that the testimony she provides is truthful. Is this not a classic expression of robustness reasoning? How else could one explain the convergence in the testimonies of the two witnesses than by assuming the reliability of the witnesses?

In response, the first point to make is that, in all likelihood, only two independent witnesses would be needed here. If there is some doubt about the reliability of the first witness, then in normal circumstances having her description of the suspect corroborated by an independent second witness should be enough to reassure us about the first witness's reliability. In other words, there is typically no need for any further witnesses to corroborate the report—the one, corroborating witness would reassure us that the original witness was not hallucinating, inventing stories, delusional and so on. If a third witness is needed, that must be because there are certain exceptional doubts about both witnesses, and I am presuming that the situation is one of normality. But if it is the case that only two witnesses are needed then we don't really have a case of robustness, since with robustness if two independent witnesses enhance the mutual reliability of the witnesses then we should expect an even greater enhancement of reliability with further corroborating witnesses. For example, with a probabilistic approach such as Bovens and Hartmann's (2003; described in chapter 1), we should expect with more independent witnesses that the posterior probability of the corroborated report would increase and eventually approach unity, based on the idea that such a convergence becomes all the more incredible the more corroborating witnesses there are. With robustness, there is no reason to expect the boon of multiple independent confirmations to elapse after a single independent confirmation. Now imagine that our police officer is a believer in robustness reasoning and that she seeks to enhance the evidential situation by retrieving testimony from as many 'independent' witnesses as possible—not with the goal of checking on potential flaws with the original one or two witnesses but simply in the hopes of creating an impressively robust, evidential scheme. As a result, she interviews 30 people who turn out to corroborate the report and then 30 more who do the same, then 30 more and so on. Is the first witness's report now approaching certainty? Leaving aside the miraculousness of having such a large number of people in a suitable position to provide worthwhile evidence reports about a crime scene, surely it is miraculous in itself that so many people would agree in their reports, given the variability in how people witness and interpret events. With such an impressive convergence, with 30, 60, 90 people unanimously agreeing in their observations, shouldn't the police officer begin to suspect some collusion occurring among the witnesses? With such profound unanimity, the hypothesis naturally arises that there is another factor motivating the convergence of reports, such as a shared societal preconception or a form of peer pressure. Sometimes observation reports can converge too extensively, a concern (recalling chapter 1) that Campbell and Fiske (1959) address with their principle of discriminant validation. In other words, achieving a broader convergence of 'independent' observation reports raises other epistemic problems, which renders doubtful the assertion that we thereby improve on the justification derived from the reports of two 'normal' observers.

We can express the reason why only two witnesses are needed in the forensics case in an alternate sort of way. With the original witness there is the possibility, we noted above, that this person is hallucinating, inventing stories, delusional or suffers from some other unusual aberration—for simplicity let us call this theoretical possibility T. If T is true, the witness's report is unreliable. Thus, to insure the reliability of the witness's report, the police officer needs to rule out the truth of T, which can be effected by securing the testimony of an independent, second witness. One is unlikely to meet two people in a row who suffer exactly the same hallucinations, narrative inventiveness and delusions; thus, should the second witness corroborate the first witness's report, we would have falsified T and established the reliability of the witness report. It is to this extent that searching for an independent observational process (such as one embodied in a second witness) is valuable when seeking to justify the reliability of an original observational process: It is a case where some theoretical possibility exists that defeats the original observational process and where another observational process has the capability of directly addressing this theoretical possibility. In this sense, it can appear that robustness is acceptable as a methodological strategy. However, strictly speaking, we are not talking about robustness here—we are talking about targeted testing. With robustness we seek independent observational evidence for a claim, that is, multiple independent processes that all attest to this claim, without regard to the details of these independent processes (except for the fact that they are independent). Apparently just by being independent and leading one to the same observed result we have a reassurance about the reliability of the processes that lead to this result by virtue simply of the 'miraculousness' of independent processes converging in this way, without needing to concern ourselves about the details of these processes. Targeted testing, in contrast identifies a weakness in the reliability of some observational process and then puts this weakness to an empirical test. Sometimes this can occur by finding a novel instance of the very same process that originally lead to the result (and whose weakness is being explored). This is what we find with the forensic witness reports described above: The second witness report effectively tests the theoretical possibility T. But targeted testing can occur in other ways. As we saw in the mesosome case, microbiologists used empirical facts to justify novel approaches to fixing microbiological specimens (such as frozen-hydration and freeze-substitution); similarly, WIMP research groups used empirical checks to ensure the accuracy of the WIMP detectors. Moreover, as we saw, empirical information can be used in a nonrobust way to calibrate an observational process, as when Perrin's authoritative determination of Avogadro's number using his vertical distribution emulsion experiments empirically tested Marian Smoluchowski's molecular theory of critical opalescence, Lord Rayleigh's molecular account of the blueness of the daytime sky as well as Planck's quantum-theoretical law of black body radiation, all theories that contained their own (subsequently corroborated) predictions for Avogadro's number. Along these lines, in our forensics case, showing that the first witness is reliable in other observational contexts could be used to show that she is reliable in the case at hand. In all these cases robustness reasoning is not occurring, even though we are utilizing alternate sources of empirical information.

Considering again the forensics case, one might suggest that two witnesses are insufficient, since these witnesses might both, and in a similar way, be disadvantaged. For example, they might have each witnessed the crime from so great a distance that they failed to notice the perpetrator's thick coat that made him look far stockier than he really is. Accordingly, the proponent of robustness might suggest, this is why we need to multiply independent observational strategies—to ensure against such misleading possibilities. We need, say, witnesses who were closer to the crime and who saw more clearly the features of the suspect, forensics experts in possession of key pieces of evidence, reports from the victims of the crime, psychological profiles of the sort of person who would perform such an act and any other (independent) piece of information that is relevant to piecing together what happened. In the end, we aim for a substantive, convergent account of the events, bound together with a full-scale robustness argument, something along the lines of 'this person is the perpetrator since, if he weren't, it would be miraculous for all these pieces of information to fit together as they do'.

But in assessing this full-scale argument, which, and how many, independent pieces of information do we need to assemble? In our original presentation of the case, two witnesses seemed to be sufficient. Now the possibility is raised, for example, that the perpetrator was too far away. In other words, another theoretical hypothesis comes to the fore, call it T′ (i.e., 'the witnesses are too far away to reliably detect the features of the perpetrator'), and, just as with T, there is need to either empirically rule out T′ or support it. So suppose we find evidence that rules out T′. Then we're back to the situation we had before, which wasn't (we argued) a robustness case but rather a case of targeted testing. Alternatively, suppose that T′ is empirically supported: Then we don't have a robustness argument either, since the testimonies of the far-away witnesses are thereby neutralized, which leaves the police officer in her report to rely solely on the testimony of any close-up witnesses. Now with the more reliable close-up witnesses, there are a variety of other forms of targeted testing that might take place. For example, perhaps there was also a tall, thin man at the scene of the crime whose presence is revealed by the more reliable, close-up witnesses. Could he have been the one who committed the crime? Here we could make recourse to video cameras, if such are available, that might contain further information about the actual event and maybe even reveal further detail about the perpetrator. Again, the strategy involves target testing the evidence produced by the close-up witnesses, showing that potential sources of error harbored by the witnesses don't apply. Or perhaps means could be put in place to calibrate the new witnesses, showing that they generate correct reports in related contexts. It is these specific demands, to target test or to calibrate, that drives the pursuit for further, independent sources of information and that sets the limit to how much, and from where, further evidence is needed. Alternatively, a blanket robustness proposal to simply find independent sources of information, regardless of a demonstrated need to address specific theoretical hypotheses, leaves the issue of testing far too open-ended. How many independent sources do we need? From what areas of research do they need to be derived? What issues should these sources of information address? Notably, where researchers are sure about the reliability of an observational process and no outstanding theoretical possibilities need to be managed, what value is there in seeking independent verification 'just for the sake of it'?

It is ultimately the silence of those who support robustness on these sorts of questions that reveals what we might call the 'excessive abstractness' of robustness reasoning. Consider, for example, the following rejoinder to how I presented the forensics case. Following the Bovens and Hartmann (2003) line of reasoning, and representing the police officer's opinion using a degree-of-belief framework, we might say that the officer's subjective probability for the hypothesis that the perpetrator was a short, stocky man does in fact increase with further independent confirmation by a third witness, a fourth witness, a fifth and so on, even if only by a very small amount—and that's enough to support the claim that it is epistemically beneficial for the officer to use robustness reasoning, leaving aside matters of targeted testing and calibration and leaving unanswered the variety of questions I posed concerning the scope and source of the independent information we are seeking. Of course, as we noted in chapter 1, the use of subjective probabilities here is problematic in that we lose the (probabilistic) independence of different witnesses. For example, upon learning the testimony of the second witness, the first witness may be emboldened in her judgment and the subjective probability of her report may increase. There is, I suggest, no reason for robustness theorists to reject this consequence—for me it simply signals the need to look elsewhere for an account of the independence of alternative physical processes than in the realm of assigning probabilities (recall that the objective probability approach had the mirror problem of rendering a witness's own reports independent of each another).

So once more, the police officer checks various witness reports and notes that the first witness's report is corroborated by a second witness report, and then she considers the value of asking yet a further witness. It may be that with the two reports, the officer is candidly convinced that the witnesses believed what they saw and is further assured that any other witness would give the same report, given a certain range in how reliable the available witnesses are expected to be. She may reflect: 'Well, that's enough witnesses—I see how this is going'. Does that mean she now assigns a probability of 1 to the accuracy of the report? Not at all—it means that she has exhausted the limits of what she may expect from the set of witnesses she is working with, leaving it open that this set is systematically biased in some respect. For instance, in the extension of the case we described above where witnesses nearer the scene of the crime are identified, the testimony of these witnesses effectively neutralizes the previous witness reports, no matter how robust these reports were originally thought to be. This is to be expected where we have a jump in the range of the reliability of the witnesses. It is precisely the sort of pattern we saw with our extended historical catalogue, where we saw scientists deferring to those observational procedures that are intrinsically more reliable. One might suggest that a scientifically inclined police officer would not only see the pointlessness of simply consulting different, though still minimally reliable witnesses: She would in fact recommend the process of targeted testing—in this case targeting the issue of witness distance as a source of inaccuracy. Or she might calibrate the witnesses, checking their vision in identifying an object with known properties; for instance, knowing that there were children playing near the scene of the crime she might ask the witnesses whether they saw them. The point is that, though multiplying 'independent angles' seems to have a sort of abstract, probative value, things look much different in real cases. What matters in real cases is finding observational procedures that enjoy an identifiable boost in reliability, which, once found, quickly usurp any purported benefit deriving from robustness arguments.

So far in this book we have been examining the issue of robustness as it applies to the empirical sciences. Still a surprising, possible source of robustness reasoning can be found in mathematical and logical reasoning. It would be an interesting and formidable result if robustness had a role to play in these central areas of scientific reasoning. My task in the next section is to consider whether robustness really does play a role in mathematics and logic.

NO ROBUSTNESS FOUND IN MATHEMATICS AND LOGIC

Suppose I am working through a long list of numbers, totaling them up. I reach the end, and I'm uncertain whether the total is right, so I tally up the numbers again, this time working backwards. The number is then corroborated. Couldn't one say that this is an excellent example of robustness reasoning? I have tried a different counting approach, and, because the result is the same, surely it must be right. The idea here is that in initially tallying up the numbers I may have been committing some unconscious error, perhaps forgetting a decimal place or double-counting some number, and one might think that in adding up the numbers again in exactly the same way I might commit the same error again. On the other hand, if I count backwards, the chances are improved that I will catch this error, revealed to me when I retrieve a different number than before. So suppose I do, in fact, count backwards on the second try and derive a different number than before. Of course it's now anyone's guess what the right answer is, so then I'm probably better just retrying the original approach, counting now much more slowly and checking to make sure I haven't made an error. Occasionally, it does indeed happen that such an error turns up. We then have a case in which both forward and backward counts (after the correction) generate the same result. Similarly, I could have derived right from the top the same number by both a forwards and backwards count. Could this convergent result be a product of an error in the original count? If that was the case, then I would be committing exactly the same error with my backwards count, and that is often too unlikely to believe. Rather, the best explanation (it is said) for why I retrieved the same number in both a forwards and backwards count must be that the count is done correctly by both methods. To paraphrase Ian Hacking (1983), it would be a preposterous coincidence to suppose that exactly the same error occurs by means of both methods.

The sort of case we are describing here is fairly common: It is any situation in which the laws of logic or mathematics are used to derive some result and in which there is some flexibility in applying these laws (such as in counting forwards or backwards, or using different electronic calculators, or having some other person perform the relevant calculation).

So let us look more closely at the case where the results of the forward and backward summing of numbers converge, abbreviating these methods as 'f-summing' and 'b-summing'. Where b-summing is found to generate the same result as f-summing, does this constitute an argument on behalf of the reliability of f-summing in the spirit of an argument from robustness? After all, this is how robustness arguments are claimed to work in the empirical sciences. Consider again Hacking's iconic example where the independent methods of electron transmission and fluorescent re-emission both reveal dense bodies in red blood cells. Hacking's tacit assumption is that this convergence establishes the mutual reliability of these methods, at least as regards the task of discerning the properties of red blood cells, and that the reality of these dense bodies is thereby established. If the convergence didn't have the effect of establishing the reliability of a process (and so of legitimizing an observational result that follows from it), it is not clear why Hacking or anyone else would have an interest in it. But if this is how we view robustness—as establishing the reliability of convergent processes—then the use of robustness in showing the reliability of f-summing is utterly inappropriate. F-summing is a reliable process, if it is a reliable process, because it is a piece of pure logic. When we learn the result of f-summing, we have learned the truth of an a priori claim. Surely it would be inappropriate to argue on the basis of an empirical inquiry that the sum of a list of numbers has a certain value, such as on the basis of the observation that f-summing and b-summing both arrive at this value. Similar comments apply to any form of logical or mathematical reasoning: Convergent proofs don't ground the claim that a form of reasoning is reliable; the reliability of a chain of logical or mathematical reasoning is inherent to the chain itself.

Another way to see this point is to consider the circumstance where, say, f-summing and b-summing arrive at divergent results. Converse robustness tells us in such a case that we should either deny the reliability of f-summing or b-summing, or deny the reliability of both methods. For instance, consider again Hacking's example where electron transmission microscopy and fluorescence microscopy both reveal the presence of dense bodies in red blood cells: If it were the case that these methods lead to divergent results, one would be forced to deny the reliability of either one of these methods, or both of them. But of course that can't be right at all in the case of f-summing and b-summing since these are both perfectly reliable, mathematical methods of reasoning. As such, where these methods arrived at different results one would not conclude that one or other of them were unreliable but would instead conclude that one or other of these methods was not, in actual fact, being used at all. In this sense, describing the convergence of mathematical or logical lines of reasoning as a form of robustness is inappropriate. Such forms of reasoning are not justified in this way.

Here one might object that the methods being used are not f-summing and b-summing in their logically pure sense but instead these methods as deployed by a fallible human agent. As such, these methods are not reliable, logically speaking, but contain a small element of human error. From here, assuming that these fallible, human forms of f-summing and b-summing are at least minimally reliable, one might suggest that a form of robustness reasoning is appropriate. Given that f-summing and b-summing arrive at the same result, the best explanation is that they each meet the logical ideal of summing—if there were sources of human error involved, such a convergence would be (as Hacking [1983] says) a 'preposterous coincidence' (201). Of course, this might not be true if the arithmetician at issue suffered from some sort of systematic counting error that showed up with both f-summing and b-summing. But leaving that possibility aside, if there is a convergence with both forms of summing, does this show the reliability of humanly fallible f- and b-summing? If this were true, then the reliability of humanly fallible summing would be an empirical matter, and just as with ideal summing this would be a misinterpretation of the reliability of humanly fallible summing. If asked, 'Why do we know that an instance of human summing is reliable?', the answer is not that this instance of human summing gives the same result as another instance of human summing. Only the most extreme conventionalist would ascribe the reliability of summing to some contingent, empirically discerned social consensus. Nor would it be appropriate to suggest that the reliability of this instance of human summing rests on the fact that it was carefully performed and free from distracting influences—these are important factors but ultimately provide no guarantee that the summing was correct, as a very poor summer could be both conscientious and distraction free. If any reason will ultimately be provided to explain the reliability of an instance of human summing, it will be that this instance of summing is performed in accordance with the logical rules of summing. So long as this is the case, the summing could have been hastily performed in the presence of multiple distractions and never reproduced by other methods—none of this would matter as regards the intrinsic reliability of the logical processes of both f- and b-summing. To further emphasize the irrelevance of robustness, suppose one arrives at the wrong result by means of a summing operation and looks for a reason for this mistake. Here one would not blame this wrong result on the fact that one had arrived at a result that was different from the results of others. In determining whether one's summing operation is either intrinsically logical or illogical, it does not matter what results other summers get—that will have no bearing on the reliability of humanly fallible summing.

Another form of 'logical' robustness involves the multiple derivation of a conclusion from a variety of starting points, what we called in chapter 1 (following Feynman; see Wimsatt 1981) a Babylonian theoretical structure. Initially this sounds like a valuable way of supporting a claim, especially if the set of starting points from which a claim is derived are exhaustive, for in such a case one can say that, whatever one believes, the claim necessarily follows. But it is also a very paradoxical way of arguing. Suppose, for instance, that there are two exhaustive theoretical alternatives, T and not-T, from which an observed result O is derivable and so predicted. Thus on the one hand, assuming T, O follows. Now suppose we assume not-T; given not-T, O follows as well. Do we now have a solid justification for (predicting) O? Consider, in such a case, the status of our initial derivation of O from T: The problem is that this derivation is completely undermined by the second derivation—if we assume not-T, it is completely irrelevant to us that O follows given T, since not-T. As an analogy, suppose one argues for the morality of a certain act A in the following way: If a deontological, nonconsequentialist ethical theory is true, then A is moral, and also if one assumes a consequentialist theory, then A is moral as well. So A, we argue, must be a moral act, since it follows whether we assume consequentialism or its opposite, deontology. But surely this is a strange way of arguing given that, if one is a consequentialist, one doesn't care at all what follows from a nonconsequentialist perspective since such a perspective will be assumed to be faulty. If one assumes that nonconsequentialism is faulty, one will likely be indifferent about the fact that from nonconsequentialism the morality of A follows, and a convergence of judgements about the morality of A from both nonconsequentialist and consequentialist positions will be thought coincidental, or at best uninformative. For instance, it may be that the morality of A is just for most people an obvious fact, and accordingly it is the duty of any theory, consequentialist or otherwise, to recapture this fact, a duty that moral theorists perfunctorily satisfy, since they must be able to handle at least the simple cases. Alternatively, each of these competing theories may independently entail the morality of A—a surprising coincidence perhaps—but that doesn't tell the proponents of either one of the theories very much because they view the theories competing with their own views as simply false. As such, they will simply ignore the claims made by competing theories and so ignore what otherwise might be thought to be robust results.

The critique of (logical) robustness we are offering here resonates with the critique Woodward (2006) offers against inferential robustness, a critique Woodward says follows the reasoning of Cartwright (1991) (see Woodward 2006, 239, footnote 13). Cartwright (1991) looks at a case in econometrics where alternative, quantitatively precise hypotheses ('functional forms') are being considered as possible representations of a fundamentally qualitative, empirical phenomenon. What econometricians do is try out different functional forms in the hopes of modeling this phenomenon, and hypothetically we are to suppose that, independent of what functional form is assumed, the same result follows. Reflecting on this case, Cartwright comments:

> [This] is the reasoning I do not understand: 'Econometrician X used a linear form, Y a log linear, Z something else; and the results are the same anyway. Since the results are... robust, there must be some truth in them.' But—on the assumption that the 'true' law really is quantitative—we know that at the very best one and only one of these assumptions can be right. We may look at thirty functional forms, but if God's function is number thirty-one, the first thirty do not teach us anything. (154)

Part of what motivates Cartwright's assessment of this (and related) cases is her belief that the functional forms conflict with each other, and only one can be accurate at any one time:

> In my diagrammatic example of functional form, we look at the phenomenon with at the very most one instrument which could be operating properly. Necessarily the other twenty-nine are bad instruments. (154)

On this sort of case, Woodward (2006) concurs. He is skeptical about the value of inferential robustness where 'a single fixed body of data... is employed and then varying assumptions are considered which are inconsistent with each other to see what follows about some result of interest under each of the assumptions' (234–235). This is precisely the sort of scenario that leaves me puzzled with logical robustness. To me, convergent inferences from conflicting assumptions amounts to not much more than a surprising but uninformative coincidence, an assessment echoed by Cartwright (1991), who explains: '[Where] all the bad instruments give qualitatively similar results' (here, the 'bad' instruments are simply those that work with conflicting assumptions) and where we have no specific argument for the descriptive accuracy of these assumptions, we are entitled to 'accept [the coincidence] just as it is, as a coincidence, or an artifact of the kind of assumptions we are in the habit of employing' (154).

Where I diverge from Cartwright (1991) and Woodward (2006) is in their contention that, whereas inferential robustness is subject to this flaw, measurement robustness is not. For both of them, the independent procedures underlying instances of robust measurements 'need not be (and it is hoped are not in fact) inconsistent with each other' (Woodward 2006, 235); rather, they only 'constitute independent instruments doing different things' and not 'different ways of doing the same thing' (Cartwright 1991, 153). Now it is doubtfully true that, in a case of inferential robustness, convergent inferences must necessarily rely on contradictory assumptions: F-summing and b-summing, for example, make use of the same set of arithmetical assumptions. Moreover, it is not exactly clear what Cartwright means when she says that observational procedures involve 'independent instruments doing different things' and not 'different ways of doing the same thing', whereas derivations from separate assumptions involve, conversely, 'different ways of doing the same thing' and not 'independent instruments doing different things'. Surely different observational procedures, if designed to generate a particular observed result (say, a value for Avogadro's number), can be said to do the same thing in different ways. Also, surely if the assumptions that ground two separate derivations of a result have nothing in common—they are 'independent'—they can be looked at as independent instruments doing different things. But the main issue for us is why Cartwright and Woodward find measurement robustness to have probative value, and here they say little except than to cite two cases: for Cartwright, the case of Perrin, and for Woodward, the case of mercury versus electrical thermometers, each case apparently illustrating how measurement robustness relies on unrelated, though consistent assumptions. Of course we are closely familiar with the Perrin case. The significance of comparing the results of mercury and electrical thermometers is uncertain without a further elaboration of the details. So, as regards measurement robustness, both Cartwright and Woodward are likely oversimplifying the scientific issues, an assessment to which our various case studies has hopefully made us sensitive.

To this point, we have argued extensively against the effectiveness, and against even the meaningfulness, of grounding the reliability of independent observational processes on their capacity to generate of robust results. But for the sake of argument, let us suppose that, nevertheless, such processes do in fact converge on the same observed result and we feel compelled to explain this convergence by means of some common cause—in other words, we take the same element of reality to be responsible for this observed result. At least in this case, do we now have an assurance of the mutual reliability of these processes on the basis of a form of robustness reasoning? I argue that we do not, for the following reasons.

ROBUSTNESS FAILS TO GROUND REPRESENTATIONAL ACCURACY

Suppose in the case just described the observed result generated by two independent processes is expressed by the sentence, 'This is an A'. We are supposing that the same element of reality is responsible for the production of this sentence in the context of each of the observational procedures. But must the element of reality that causes the independent production of the report 'This is an A' be itself an A? Indeed, must As exist at all, despite the eponymous reports? It is easy to imagine instances where this is not the case. Consider again Locke's fire example. Suppose an observer thinks that fire is actually caloric—heat substance as understood by 18th C chemistry. As such, whenever this person sees a fire he utters the report, 'Caloric!' Now suppose further that whenever he sees caloric at a distance and feels uncertain about whether he might be hallucinating, he reaches out his hand to determine whether he can also feel the heat of the caloric, and when he does, again utters, 'Caloric!' Does the robustness of this observational report, as generated by two independent observational procedures, enhance the reliability of his observation report? Obviously not, since there is nothing in the world that fits his description of what is being called 'caloric'. Moreover, there is nothing in the practice of robustness itself that could expose this flaw. What exposes this flaw is a direct reflection on the reliability of the observational process that leads up to the utterance, 'Caloric!' Notably, one reflects on the category 'caloric' and considers the empirical evidence at hand relating to whether such a substance really exists, perhaps taking into account the pivotal empirical researches of Count Rumford that disprove the existence of caloric. Given what we know now about heat phenomena, we judge any observational process culminating in the report 'Caloric!' to be unreliable since it incorporates an inaccurate categorization.

Here the case involving mesosomes is similarly instructive. It was noted that, if robustness were the chosen strategy of experimental microbiologists, their conclusion would have been that mesosomes exist: Non-observations of mesosomes occurred under relatively special conditions, that is, in the absence of prefixatives, fixatives and cryoprotectants, whereas observations of mesosomes occurred under a variety of circumstances. Thus, one might argue in accordance with robustness that there is some element of reality that causes the consistent observation of mesosomes—but is this element of reality some native feature of the substructure of bacteria, a sort of organelle with a unique function? Many microbiologists believed this to be the case, and though they were wrong about what element of reality they thought they were observing, they were at least right that there is an element of reality that causes their robust observations. It just turns out that this element of reality is somewhat different from what they expected—that is, it is actually an artifact of the preparative process for bacteria. This fact was discovered by various empirical inquiries revealing the distortions caused by the use of OsO_4 and other fixative agents, inquiries that show the non-naturalness of the mesosome category, the robustness of observations apparently revealing their existence notwithstanding.

Another way to see how the robustness of an observation report has no necessary link with the representational accuracy of the report is to consider the evidence for the existence for dark matter available prior to the discovery of the Bullet Cluster. There was, we noted, empirical evidence for the existence of dark matter from the rotation curves of spiral galaxies, the velocity distributions of galaxy clusters and gravitational lensing. But such robust evidence can be used to support a competing theoretical picture—a modified gravity approach, such as MOND. In other words, there is nothing in robustness that solves the underdetermination problem concerning these two competing theoretical representations of reality. One must step outside robustness and use a different strategy to handle such underdetermination problems (such as using what I called 'targeted testing') so as to be more precise about which theoretical viewpoint is best supported by the empirical evidence. In other words, robustness may inform us that there is some element of reality that is causally responsible for a set of robust results, but it doesn't have the resources to tell how best to describe this element of realty.

Perrin's various determinations of Avogadro's number raise another problem for the issue of the representational accuracy of robust observations. Perrin describes various methods for arriving at Avogadro's number. I questioned whether Perrin's reasoning was truly robust (it turned out to be more of a calibration). But leaving that exegetical matter aside, and supposing that his argument was indeed based on robustness reasoning, we noted that Perrin's estimation of Avogadro's number, from a modern perspective, was rather imprecise and strictly speaking inaccurate. Of course, the response often given here is that it is remarkably close to the appropriate order of magnitude we need to be working with—but I noted that this assessment is not without controversy. The key point for us is that, even in a case where there is an element of representational accuracy (albeit rough), robustness does not contain the resources to improve on this accuracy. Rather one improves on this accuracy by setting up observational procedures that are theoretically designed to be more reliable indicators of Avogardo's number, such as with the recent use of the XRCD method, which measures *N* close to eight decimal places (see, e.g., Mohr et al., 2008), by comparison to Perrin's determination to one or two decimal places.

THE SOCIOLOGICAL DIMENSION OF ROBUSTNESS

Though I have argued that robustness lacks the epistemic value many have ascribed to it, it is nevertheless true that some scientists portray themselves in their philosophical moments as utilizing such reasoning (Kirshner and Perrin are two cases in point), and that many (if not most) philosophers regard robustness as one of the prime strategies for ensuring the accuracy of observational data. It would therefore be valuable to have an explanation for this support, which I believe is forthcoming from sociology.

In all the cases we have been examining, the social contexts in which the scientists are working are disputational in that scientists are challenged to provide justifications for their beliefs in the face of empirical or theoretical challenges put forward by scientific competitors. Whether it be mesosomes, WIMPS, atoms, dark matter or dark energy, the proponents of the existence of these things encounter profound and dedicated criticism and are forced to diligently defend themselves. Now what I claim our case studies tell us is that scientists strive to address this disputational environment by seeking to improve the reliability of their observational procedures. This is to me a rational way to proceed in managing these disputational pressures, one that can be enhanced through the additional strategies of targeted testing and calibration.

Still, it can happen that a scientist is pressured by a disputational situation to find a justification for her observed results that extends beyond what the basic empirical findings tell her. This may happen, for example, when the empirical findings are inconclusive but there is a need to firmly justify a result (perhaps to convince students in a pedagogical situation or in a popular context to convince a wider, nonspecialist audience). Where such pressure exists, what more can the scientist suggest in defense of her results? This is where a generalized strategy such as robustness can serve an invaluable purpose, for it holds the key to a unique argumentative strategy that can provide a new line of evidence against one's detractors. It works in this way because it references alternative observational strategies (meeting a minimal reliability requirement) whose characteristic feature is that they are independent of the original strategy—without needing to say how exactly these strategies differ. Consider again Hacking's (1983) iconic example, where 'two physical processes—electron transmission and fluorescent re-emission—are used to detect [dense bodies in red blood cells]' (201), and let's suppose that fluorescent re-emission couldn't be used but that there was some other method that could be used and that would give the same observed result. For robustness to work, it really doesn't matter what this independent alternative method is, so long as the minimal reliability standard is met. If perchance palm reading meets this standard, then palm reading could be used as an alternative method for the purposes of robustness reasoning. In other words, an interesting feature of reasoning robustly is that one need not have any knowledge whatsoever of how an alternate observational procedure works, since for robustness to work one need only know that an alternate procedure is minimally reliable and independent of one's original procedure. The scientist, then, under pressure to defend her views beyond what her basic findings suggest, has a potentially large resource of robust data with which to work, data that is effective even if she is unable to give the details underlying this effectiveness. It's analogous to having at hand a whole new world of evidence for one's views without needing to bother with the details for why, precisely, this evidence works. As an extra bonus, it's evidence that even nonscientists can appreciate since they, too, don't need to know the exact scientific details underlying an alternate observational procedure, only that this procedure is minimally reliable and suitably 'independent'. So where there's pressure to defend one's results to, in particular, nonscientists, robustness reasoning can be quite useful.

The usefulness of robustness reasoning, as we have described it, is not limited to referencing inanimate observational procedures. Consider again a case in which a scientist arrives at an observed result the justification of which is subject to dispute but in which the extant evidence is ambiguous. Where there is pressure to resolve the issue, the scientist has the option of calling on an impartial and supportive third party to intervene who, if authoritative, can act as an effective independent locus of support. Assuming the third party is at least minimally reliable, the independent testimony of this individual can provide the basis for a robustness argument that can (purportedly) enhance the quality of the evidence. No doubt, many debates in the sciences and in other intellectual areas follow this dynamic, where (independent) authorities step in and (at least temporarily) resolve intellectual disputes simply by virtue of their presumed independence. The particular convenience of this strategy is its low threshold: So long as the third-party interveners meet the minimal reliability and independence requirements, no one need know anything further about the details of the authority's line of reasoning. We are simply left with the surprise of the convergent opinion, best explained by the truth of the observed result, and robustness does the rest. It is critical though that we recognize the epistemically limited nature of these third-party authoritative interventions, despite their social benefits in managing intellectual controversies. For instance, it is perhaps such an allusion to authority that Kirshner found useful in conveying to a popular audience the accuracy of his research group's observation of the universe's accelerative expansion. But when it came to a matter of recapitulating, in the context of a Nobel Prize lecture, the crucial reasoning on behalf of such an expansion, the representatives of both SCP (Saul Perlmutter) and HZT (Brian Schmidt and Adam Riess) neglected to mention the surprising convergence of their views. If indeed robustness reasoning has the 'ring of truth', as Kirshner (2004) suggests, one would have expected this convergence to have been front and centre in a Nobel Prize lecture. The point is that the particular merit of robustness reasoning—that it is compelling even if one lacks a detailed understanding of the (minimally reliable) observational processes at hand—is at once its main drawback: When asked why an observational process is reliable, a scientist will need to do much better than simply cite the convergence of this process's results with its results with another (minimally reliable) observational procedure.

Chapter 7

Robustness and Scientific Realism

So far we have been examining and questioning the value of robust observational procedures. There are, however, other sorts of information-gathering procedures that could be said to be robust. In chapter 6 we examined robustness reasoning in the context of mathematics and logic, where trains of independent yet analogous forms of reasoning lead to identical conclusions. Similarly, one could use robustness reasoning to argue against ethical relativism. For instance, in examining independent cultural belief systems, one might note how people in each of these systems advocate the same fundamental moral principles, despite having never interacted (e.g., one might observe that people in different cultures independently converge in their condemnation of cold-blooded murder). Given this convergence of moral opinion, one might infer the (a priori) truth of the relevant moral principles. In the spirit of locating such varied instantiations of robustness reasoning, I consider in this chapter a form of robustness reasoning that, I believe, has a place in the thinking of many philosophers of science, a form of reasoning that plays a key role in the defense of scientific realism. On this approach, it is noted that different scientific theories in the past have been found to express theoretical claims that reappear in subsequent, sometimes conflicting theoretical settings. In other words, such claims are robustly generated, reproducible in independent contexts, which for some realists is an indicator that these claims have a special epistemic status. Thus, robustness reasoning is found to make a surprise appearance in the philosophical defense of scientific realism, and, as the reader might suspect given my skeptical view of the value of robustness, I do not view such defenses of realism to be promising. In what follows I illustrate more fully how robustness reasoning plays a role in arguments on behalf of scientific realism, and from there proceed to critique this application of robustness by reference to the historical case studies examined in this book. In due course I propose a different approach to defending realism that avoids robustness (called 'methodological preservationism' to contrast it with the 'theoretical preservationism' favoured by many contemporary realists), an approach that is itself illustrated and motivated by these same case studies.

To get started in understanding why scientific realists have felt compelled to adopt a version of robustness reasoning, let us consider some of the philosophical background related to arguments for and against scientific realism.

THE NO-MIRACLES ARGUMENT FOR SCIENTIFIC REALISM

Scientific realism claims that our best, current scientific theories are at least approximately true descriptions of the world, and the current, main argument in support of scientific realism is the so-called 'no-miracles argument'. According to this argument, if our best, current scientific theories were not at least approximately true, then it would be miraculous for these scientific theories to be as successful as they are. Conversely, the main argument against scientific realism is based on what is called the 'pessimistic (meta-)induction'. This argument starts with the observation that what counted in the past as our best scientific theories often turned out to be false as science progressed. Famous examples of this tendency include Newtonian mechanics and Maxwell's ethereal theory of electromagnetism, both of which were falsified by Einsteinian relativity theory. The lesson from these episodes is that we should be wary of our current theories for, despite their success, odds are that they will themselves be rejected by later scientists, the no-miracles argument notwithstanding. A related argument against scientific realism is the 'underdetermination argument'. Given any (successful) scientific theory, an empirically equivalent though logically incompatible theory can be constructed (perhaps very artificially), and so the empirical support we have for our current, best scientific theory is ultimately equivocal—it could just as well provide support for a competing, incompatible theory, a competing theory that moreover could be the beneficiary of an analogous no-miracles argument. Stanford

(2006) has questioned the force of the underdetermination argument on the basis of his incredulity about the possibility of meaningfully constructing empirically equivalent alternatives to our best theories. In its place he advocates his 'new induction' based on (what he calls) the 'problem of unconceived alternatives': As Stanford suggests, for any scientific theory in the past there have been (logically incompatible) subsequent theories that just as well capture the empirical evidence that the former theory captures but that were unconceived (or even unconceivable) for the proponents of the original theory. As a result we should once more be wary of our current theories because, despite their empirical success, odds are there are logically incompatible theories that will be formulated later on that will be equally well (or even better) supported by the same evidence.

There are a variety of ways by which a realist can rebut the pessimistic induction (and the related problems of underdetermination and unconceived alternatives). The most common is to adopt a form of 'preservationism', or what I more perspicuously call 'theoretical preservationism'. On this approach, past successful theories that are subsequently claimed to be false are analyzed in a way that separates out those parts of the theories that, from the perspective of hindsight, can nevertheless be asserted to be true. Two examples of such a strategy involve (a) the caloric (or fluid) theory of heat, which was subsequently replaced by a 'molecular motion' theory; and (b) Maxwell's ethereal theory of electromagnetism, replaced later on by Einstein's nonethereal theory. As regards the former, Psillos (1994) and Psillos (1999) argue that the successes of caloric theory are explicable without reference to those parts of the caloric theory that were subsequently rejected—that is, in Hasok Chang's (2003) paraphrase of Psillos's views, we retain 'the laws of calorimetry, the adiabatic law and Carnot's theory of heat engines' in the molecular theory (904) but dispense with any reference to the existence of caloric itself. Philip Kitcher (1993) gives a similar assessment of Maxwell's theory of electromagnetism: The working core of Maxwell's theory (his four equations) was retained and used in explaining electromagnetic phenomena, while Maxwell's postulation of ether serving as the medium of wave propagation was dispensed with. This strategy, called by Psillos (1999) the 'divide et impera' move, saves the no-miracles argument by restricting the successful parts of past theories to those parts that really and accurately refer to entities in the world, at least from the perspective of more current scientific theorizing. Those parts of past theories that are 'preserved' in current theory are said to have been responsible for the successes of past theories and to also explain the analogous successes of new theories. The pessimistic induction is thus defeated by rejecting its premise: When we restrict ourselves to the preserved core of a theory, the success of a theory, wherever it occurs, can be explained by reference to this core, as this core is not subsequently falsified.

Theoretical preservationism has become very popular as a rejoinder to the problems facing scientific realism. One of its most developed forms is 'structural realism', which identifies in theory change the preservation over time of theoretical (often mathematical) 'structure'. Here we attempt to understand why preservationism is so popular, drawing initially from the work of one of the main proponents of structural realism, John Worrall.

IN SUPPORT OF THEORETICAL PRESERVATIONISM

In the face of the pessimistic induction, Worrall (2007) argues for preservationism (or more specifically, structural realism) in the following way:

> It is of course logically possible that although all previous theories were false, our current theories happen to be true. But to believe that we have good grounds to think that this possibility may be actualized is surely an act of *desperation*.... Any [such] form of realism seems *patently untenable*. Only the *most heroic head-in-the-sander* could... hold that our current theories can reasonably be thought of as true [given the pessimistic induction].... [Believing this] would be a matter of *pure, a-rational faith*. (129–130; my italics)

Thus, to be a realist on Worrall's view, one must suppose that previous theories were not entirely false, that at least the successful ones were correct about the '"deep structure" of the universe' (133). That is, it must be the case that past scientists got some claims right (for Worrall, at least about the 'structure' of the world) and that some of these claims are 'preserved' (as true) in our present-day science, for otherwise we would be forced to conclude with the pessimistic induction that scientists could never get anything right at all.

Unfortunately the argument Worrall is providing here for preservationism is riddled with ad hominems; even if nonpreservationists are desperate, a-rational head-in-the-sanders, that says nothing about the doctrine of nonpreservationism itself. He provides a better form of reasoning in a footnote. First of all, he acknowledges that scientific theories are improving: 'Later theories are better empirically supported than their predecessors' (129, footnote 7). But on his view the fact that later theories are better supported than earlier ones does not imply that later theories will not, themselves, be subsequently replaced and found to be false by the lights of an even later theory. Why not? To accept such an implication would be analogous to suggesting that 'the current 100m sprint record will [not] eventually be broken' because 'the current [100m sprint] record is better than the earlier ones' (130, footnote 7). Here, Worrall's reasoning seems forceful: Just because science has improved doesn't imply that it cannot be improved further, which is to say that just because a current scientific theory has been asserted to be true on the basis of improved grounds (in comparison to past theories that have correlatively been found to be false), that doesn't imply that it won't be found to be false later on the basis of yet further, improved grounds. Accordingly, there is no bypassing the pessimistic induction by making reference to improved standards: Even with improving standards, once past theories have been found false one can induce that future theories will be found false too.

Once again, the preservationist response to this challenge is to deny the premise that past theories have (in their entirety) been found to be false. The belief is that there are preserved parts that were truthful in the past and truthful in the present. The argument for this belief is that these preserved parts must exist, or else we would have no grounds in the least for asserting the truthfulness of our current theories.

The issue of improving standards in science is a key one, as I argue below, and provides the framework for a realist rebuttal to the pessimistic induction without making recourse to (theoretical) preservationism. However, it is inaccurate to suggest that the standards in science will be improved indefinitely. Here, the sprint race example is apt. Suppose that the current 100m sprint record is x and that this record is the product of a long series of year by year, marginal improvements that have run their course to a maximum. Humans, let's suppose, have reached their pinnacle in this regard, so much so that it's hard to see how any human could improve on this record. Under these circumstances, one can, contra Worrall, draw the inference that the current 100m record will stand its ground, precisely because it is an improvement over past records (so long as we add in that the record of x has not been improved on for a long time and that we have trouble even seeing how it could be improved further).

But before we turn to the issue of standards, let us examine one further argument for preservationism, an argument that bears a strong resemblance to a form of robustness reasoning. Consider again the caloric theory of heat and Maxwell's theory of electromagnetism. According to preservationism, each of these theories has components that are preserved in later theories; for example, the laws of calorimetry are preserved in modern theories of heat, and Maxwell's equations are retained in modern-day electromagnetism. What might be thought somewhat amazing is that these theories succeeded in generating successful, and subsequently preserved, components, despite their allegiances to faulty ontologies. How can reflecting on heat substance and the ethereal medium generate accurate calorimetric and electromagnetic laws? To some philosophers, the fact that caloric theorists Joseph Black and Antoine Lavoisier (see Chang 2003) and ether theorist Maxwell (see Stanford 2003 and Stanford 2006) needed to invoke caloric and ether, respectively, in their theoretical derivations works against the preservationist rejoinder to the pessimistic induction. The reason is that the hypotheses of caloric and ether are, as a consequence, in part responsible for the successes of theories of which they are a part; thus, there is no dismissing them in explaining these successes (see Doppelt 2007 for further reasoning along these lines). In other words, in just focusing on the preserved parts of these theories (which preservationists tend to do), we lose the explanatory and empirical successes of these theories and so lose what it is the no-miracles argument is meant to explain.

But there's another way we can look at the need to retain subsequently rejected theoretical components in accounting for the explanatory/empirical success of past theories, and that is to view past theories and present theories as simply different strategies at generating the same successes. For example, given the hypothesis of caloric, past scientists generated the laws of calorimetry, and, today, without the hypothesis of caloric, scientists are again also able to arrive at the laws of calorimetry. Similarly, given the hypothesis of ether, Maxwell generated his namesake laws; today, without the hypothesis of ether, scientists are able to arrive at Maxwell's laws. Now we can extend this strategy to cover other theories that historically intervene between the past theory and the present theory. Each intervening theory is distinctive in what assumptions it takes to be true, and, supposing it is successful in preserving the same elements that are preserved in present-day theories (such as the calorimetric laws or Maxwell's equations), we have yet another example of how from differing assumptions the same true, preserved results follow (whether or not these assumptions are, in fact, true). My suggestion, accordingly, is that we can locate at the theoretical level a form of robustness reasoning that can be used to support the preserved elements of theories: Just as empirical claims are purportedly vindicated by having been generated through differing experimental strategies, so are theoretical claims purportedly vindicated by having been generated through differing theoretical derivations.

This theoretical version of robustness has, I believe, wide application and wide appeal. Think of when theoretical claims have been said to 'pass the test of time'. Some of these are moral claims—for example, when a controversial decision made by some political leader has been 'vindicated by history'; some are aesthetic claims, such as when the value of an artwork has proved its mettle over the years; some are philosophical claims—the inherent value of the Platonic dialogues is shown by the fact that philosophers continually return to them in their teaching and research. The argument then runs as follows: People of different eras, cultures and intellectual backdrops have found value in this politician's decision, this artwork, this philosophy; thus, these objects of value reveal something important, such as a deep truth or insight—for how else can one explain this convergence over time? Surely, it is argued, this convergence cannot be explained by the idiosyncratic nature of some culture, era or intellectual backdrop, since there is agreement in these value judgments despite differences in culture, era or background. A similar sort of argument may arise in the justification of a democratic mode of governance. How do we know that a democratically elected leader is the best for the job? Supposing for simplicity that the leader gained a substantive majority, the argument is that the leader received the votes of people who come from a variety of age groups, economic classes, religious backgrounds, political affiliations and so on, so it cannot be simply that this leader is the pet favorite of some interest group; rather, some other quality of the leader must explain this success, specifically, the fact that he or she is the best candidate for the job.

My suggestion then is that we can find support for theoretical preservationism in a form of robustness reasoning, here applied at the theoretical level. What we find is that robustness not only plays a role in a prevalent understanding of how observational practice can be reliable but also plays a role in a prevalent understanding of how scientific realism can be maintained in the face of a history of (apparently) successful but ultimately false theories. The idea is to identify preserved elements of (successful) theories that are common to scientists working in different eras, cultures and intellectual backdrops and to assert that we can reliably support the reality of these elements solely in light of their preserved status, even if we find ourselves unable to support the other parts of these theories that have a more restricted range. As such we can say that preservationism benefits from a form of 'theoretical' robustness reasoning.

Now if robustness is indeed being applied at this theoretical level, then one would expect that the critiques I have launched against robustness in the area of scientific observation could apply as well to robustness found in the study of scientific historical episodes. Indeed this is what we find: Some recent criticisms of preservationism in the literature are remarkably similar to some of the critiques I launched against robustness.

OBJECTIONS TO THEORETICAL PRESERVATIONISM

Recall that one worry with robustness reasoning is the question of how we can be sure that diverse observational approaches to confirming an empirical claim are genuinely independent. It is just this sort of concern that animates Stanford's (2003) critique of preservationism. Preservationism, he says,

> faces a crucial unrecognized problem: of any past successful theory the [preservationist] asks, 'What parts of it were true?' and 'What parts were responsible for its success?', but both questions are answered by appeal to our own present theoretical beliefs about the world. That is, one and the same present theory is used both as the standard to which components of a past theory must correspond in order to be judged true and to decide which of that theory's features or components enabled it to be successful. With this strategy of analysis, an impressive retrospective convergence between judgments of the sources of a past theory's success and the things it 'got right' about the world is virtually guaranteed: it is the very fact that some features of a past theory survive in our present account of nature that leads the realist both to regard them as true and to believe that they were the sources of the rejected theory's success or effectiveness. So the apparent convergence of truth and the sources of success in past theories is easily explained by the simple fact that both kinds of retrospective judgments about these matters have a common source in our present beliefs about nature. (914; see also Stanford 2006, 166–168)

I quote Stanford at length because this is exactly the sort of concern we should have with robustness when applied to the validation of any empirical claim. If we have already settled on which empirical claim needs supporting, then it is a relatively simple matter to find diverse observational strategies that 'converge' in support of this claim: Any observational strategy (meeting a minimal reliability standard) that issues in this claim we deem 'successful', and strategies that fail to generate this result we either ignore or dismiss for spurious reasons as 'unreliable'. On this basis we argue robustly that the claim is likely true. A similar surprising source for this worry derives from Orzack and Sober (1993) in their discussion of the robustness of models. Sober, who (as we saw) is otherwise a supporter of robustness, considers the required degree of independence needed for robust modelling to be 'unfortunately...elusive' (Orzack and Sober 1993, 540). In this vein, Orzack and Sober recommend that we exercise care in

> considering the possibility that robustness simply reflects something common among the [intellectual] frameworks and not something about the world those frameworks seek to describe. (539)

This is precisely the problem that Stanford claims we will find afflicting the empirical support of theories when present-day theorists look to past theories to find a convergence on the 'true view'; such theorists are said to be committing the intellectual flaw called 'presentism' or 'Whiggism', judging the past on the basis of the present. Chang (2003) shares a similar worry; with regard to what he calls 'the most fundamental problem with preservative realism', he says,

> Even when we do have preservation, what we are allowed to infer from it is not clear at all. The uncertainty arises from the fact that there are several different reasons for which elements of scientific knowledge may be preserved. Beliefs or practices may be preserved either because nature continually speaks in favor of them, or because our own cognitive limitations confine us to them, or because we just want to keep them. The inference from preservation to truth can be valid only if the latter two possibilities can be ruled out. Even extraordinary cases of preservation, in themselves, do not necessarily show anything beyond human limitations, or conservatism assisted by enough obstinacy and ingenuity. Preservation is far from a sufficient condition for realist acceptance. (911–912)

This is the exact analogue to the sort of problem we can find with robust empirical results. For instance, it might turn out that various observational strategies are found to lead to the same observed result because we lack the cognitive capacity to think of strategies such that, were they instantiated, would lead to different results. Or perhaps we have a bias toward a certain observed result that leads us to dismiss (as 'unreliable') observational procedures that don't 'cooperate'.

There is reason, then, to think that the various arguments I have provided in this book against robustness reasoning as applied to observational processes can be analogously marshaled against preservationism, insofar as preservationism is motivated by a form of robustness reasoning that identifies common elements in a series of past successful, though largely discarded theories. Consider, for example, the claim we saw above, that both caloric theory and molecular motion theory can generate the laws of calorimetry and that both Maxwell's ethereal theory and Einstein's nonethereal theory can generate Maxwell's equations. In other words, the laws of calorimetry and Maxwell's equations are 'preserved', generated, respectively, by an older theoretical perspective and a newer one, and so by a preservationist 'robustness' argument one is in a position to be realist about these laws and equations. Of course, I suggested (in chapter 1) that robustness is not a valuable approach when we are considering two observational procedures, one of which is deemed reliable and the other unreliable. What value is there, one might suggest, in considering the testimony of an unreliable observational strategy when one has at hand a reliable observational strategy? Analogously, one might argue, why bother considering the testimony of an unreliable theoretical perspective (such as caloric theory or ether theory) when deciding on the truthfulness of a result derivable from a more reliable theoretical perspective (such as the molecular motion theory or Einstein's theory of relativity)? For this reason, one might feel inclined to question the authority of a preservationist argument for realism.

However, my plan now is to let this concern pass: Instead of reiterating my previous arguments against the epistemic significance of robustness as applied to observational processes and then directing these arguments in analogous fashion to the case of preservationism, my plan alternatively is to address the case of (theoretical) preservationism directly to see whether it has force in grounding a realist interpretation of theories. For instance, Stanford (2003, 2006) and Chang (2003) have revealed some reasons to doubt the force of preservationism, first where there is a lack of independence in determining what elements are preserved across theory change (Stanford), and second where the preserved elements are identified for reasons that are arguably nonepistemic (Chang). My plan is to further their critiques of preservationism, and derivatively to further my critique of robustness, by arguing on historical grounds that scientists are inclined to rebuff preservationist considerations in their empirical inquiries (here distinguished from purely theoretical inquiries where, in the absence of new and possibly anomalous observational data, preservationism is a much easier doctrine to support). The overarching idea is that if scientists can be found to ignore matters of preservation—that is, if they tend to avoid the task of accommodating and maintaining past theories—then we have found yet another reason why we should deny the force of robustness, generally speaking. This is because preservationism and robustness share a similar logic; to wit, they both assume that, in having independent routes to the same conclusion, we thereby put this conclusion on firmer epistemic footing. I propose to show in historical terms that empirical scientists eschew theoretical preservationism and so eschew the underlying logic of robustness.

The historical inquiry I undertake to this end need not take us far from the case studies we have already examined in this book (i.e., concerning mesosomes, WIMPs, atoms, dark matter and dark energy). We find that in none of these cases does preservationism impose a constraint on how scientists assess their results. In fact, if scientists were really committed to preserving what were previous theoretical insights, then the pivotal theoretical discoveries these scientists think of themselves as having made would have never happened. It would clearly be a negative feature of preservationism if it not only failed as an interpretation of past practice but also had the residual problem of cramping scientific discovery.

Before turning to these cases, however, we need to affirm the following caveat. By denying the value of preservationism, we are not denying that scientific advance characteristically builds on past theoretical developments; we are not denying that new scientific theories take for granted an enormous body of accepted methodological and doctrinal background. This is, of course, true—it would be extraordinarily counterproductive for scientists to continually begin from square one. Rather, the point is that the preservationist argument—namely, that if a theoretical (or structural, or phenomenological) claim is preserved in a later theory, then these are grounds to view this claim as true, or what it describes as real—is a bad argument and that philosophers should not be advancing such an argument in defense of a realist interpretation of scientific advance, just as no self-respecting scientist would ever make this claim in suggesting a reason to support a purported empirical discovery. In other words, it really doesn't make much sense for a scientist, aware of the enormous shifts in scientific theorizing over time, to attempt to preserve an ontological viewpoint for the simple reason that preservation has some purported but unspecified special value. Scientific claims should be preserved for the right reasons, such as their justification on the basis of reliable empirical results. But if this is correct, then philosophers are doing the progress of science a disservice in suggesting that scientific realism can be defended and that the pessimistic induction can be denuded by advocating theoretical preservationism. Preservation for preservation's sake is no sign of truth, whether understood prospectively in the scientist's hands or retrospectively in the mind of a philosopher.

With these comments in mind, let us return to our case studies.

The mesosome. The mesosome, when first discovered, was a brand-new theoretical entity that was completely unexpected for microbiologists at the time. Scientists argued for the reality of mesosomes using what I termed reliable process reasoning, whereby the relevant reliable process involved the use of the Ryter–Kellenberger (R–K) fixation method, viewed in the 1950s as the standard method by which microbiological samples were prepared for microscopic investigation. Later on, when it was realized that the process of osmium fixation (intrinsic to the R–K method) could be creating mesosomes, experimenters had to revisit the claim that the R–K method was a reliable approach, and eventually they conceded its tendency to create artifacts. From the perspective of preservationism, what needs to be emphasized here is that, with the (purported) discovery of mesosomes, the former view of bacterial substructure (that it is 'organelle-less') was displaced, not preserved. The organelle-less view had, until the discovery of the mesosome, been theoretically entrenched, vindicated repeatedly by the observations generated by pre-electron microscopic technology. Then, after the reality of the mesosome became (for a short while) the new theoretical norm (having been purportedly established by electron microscopic observations), there was no movement on behalf of microbiologists to resist its inclusion in the new microbiological ontology given the fact that it didn't 'preserve' the old, organelle-less view. Adopting a preservationist viewpoint along these lines wouldn't have made any sense: A new observational standard had been set with the introduction of an electron microscope, a standard that was for all concerned a clear methodological improvement, and it would have been simply backwards to turn a blind eye to what the new microscope revealed. Here, one can even view the situation in a sort of structuralist way—the old perspective of bacterial substructure, as organelle-less, was displaced by a new form of substructure. To intentionally preserve the old substructure in one's theoretical understanding of bacterial morphology, where a new form of substructure is clearly empirically supported, does not (and did not) make any scientific sense.

Of course, as the story continues, mesosomes were discovered to be artifactual, which meant a return, to some extent, to the pre-electron microscopic view of an organelle-less substructure. But no one argued for this conceptual reversion on the basis of some sort of preservationist impulse; no one reasoned that, once the mesosome was set aside and we found a convergence of views linking pre-electron microscopic observations and the new, post-R–K methodology, electron microscopic observations, there was an extra merit to be attached to the no-mesosome result because it 'preserved' a prior viewpoint. Reasoning in this way would be pointless because the earlier nonorganelle perspective was based on impoverished light-microscopic evidence, and there is no expectation that this evidence would be particularly reliable by comparison to the new methods that had subsequently been invented.

It's worthwhile pointing out here that sometimes alternate empirical methods were used in the mesosome case to address issues with the primary experimental methodology, methods that often focused on possible sources of experimental error. As a case in point recall that Silva et al. (1976) and Hobot et al. (1985), in asserting that mesosomes are artifactual since they are generated by experimental methods that employ OsO_4 fixation, justify their suspicions that OsO_4 fixation damages cell structure by pointing to other experiments that test and confirm this claim (as we saw, Silva et al. [1976] highlight the fact that OsO_4 damages the permeability of the cytoplasmic membranes of bacteria, and Hobot et al. [1985] point to its ability to rearrange cellular content). These sorts of circumstances illustrate the sort of exception we have allowed in our contra-robustness argumentation—that targeted, alternate observational strategies are a perfectly legitimate way to ensure the reliability of a primary empirical technique. Here, these extra empirical methods serve to test the judgment that OsO_4 fixation leads to mesosomes: The more tendentious result—that there is no proof for the existence of mesosomes because OsO_4 fixation is not reliable—is vindicated by introducing straightforward, observational examples where the flaws of OsO_4 fixation become apparent. In this situation, independent empirical data are valuable not because they generate convergent observational claims but because they confirm an assumption that is fundamental to the generation of a certain, key observational report (here, that no mesosomes are observed with an improved experimental regimen).

The WIMP. Theoretically, the WIMP is not an absolute newcomer, for it is understood to be represented theoretically as a neutralino, a representative part of the supersymmetric extension of the standard model of particle physics. But the neutralino is in no way an established particle; it is hypothetical, and discovering WIMPs could in turn provide support for the existence of neutralinos. So in asserting that WIMPs exist, DAMA is not in any way preserving an established theoretical viewpoint, a set of phenomenological laws or any amount of empirical data—its assertion involves a quite new entity with almost no scientific lineage.

It is of course true that DAMA's claim to have observed WIMPs came under fire from a number of teams of astrophysicists. But no team ever argued that the WIMP, as a completely novel entity that failed to preserve some prior theoretical viewpoint, was therefore a questionable entity. Arguing that way wouldn't make much sense—how could it be reasonable to generate novel observational results if one felt obliged to toe some line that demanded the preservation of a prior theoretical perspective, one that, in the case at hand, would have to exclude WIMPs? By definition, a novel observational result is one that fails to preserve some aspect of prior theorization.

It's worthwhile emphasizing what's at stake here. Generally speaking, the value of novel advances shows us why preservationism is leading us in the wrong direction. Where we are dealing with a novel advance, the whole point of the advance is to introduce the existence of something (a law, an entity, a kind of entity) that has not been conceived of before, something that could not be the subject of preservation because it was not available for preservation in the minds of scientists. Seen in this way, preservationism is a counsel for conservativism where novel advances are resisted and traditional approaches upheld for the sake of their traditionalness. Probably the main candidate for a conservative view of science is a form of common-sense realism—and surely such a realism would spell disaster for scientific progress.

At this stage, one might defend preservationism by noting that it doesn't discourage the search for novel facts but only emphasizes the value of theoretical conceptions that have stood the test of time. For the sake of argument, let us suppose that novel advances don't conflict with established theory—they simply populate the theoretical world with something new. Note however that, if we are at all convinced by the pessimistic meta-induction, we should be convinced as well by its application to novel facts, for no doubt the history of science is filled with cases where a novel phenomenon has been witnessed or a novel theory had been introduced, and such novel advances were later repealed in subsequent scientific developments. Moreover, for such advances there is no recourse to a preservationist rejoinder to the meta-induction since, as novel advances, there is nothing to preserve. Now consider the case where a novel advance conflicts with antecedently held, long-standing and well-preserved theories. Such novel advances are clearly flying in the face of what we should believe, according to preservationism. I contend in fact that this is a very common situation in the history of science—practically any case that we could describe as a paradigm shift involves the rejection of relatively persistent, established theoretical assumptions in deference to some brand-new conception at odds with the old paradigm. It follows then that preservationism counsels us to avoid the sorts of radical conceptual changes found in paradigm shifts, and it is hard to see the epistemic value in such a recommendation.

Atoms. The atomic theory of matter has been around for eons but did not become the dominant theory of matter until the early 20th century. When Jean Perrin was arguing for the reality of atoms, there were many other scientists who were prepared to assume the existence of atoms. Nevertheless, Perrin's task (following Einstein) was to respond to the proponents of classical thermodynamics who still resisted the atomic hypothesis. We saw how he went about achieving this task, and in no sense did it involve a preservationist type of argument, one where Perrin noted the experimental researches of others and argued that because they arrived at the conclusion that atoms exist, his own conclusion that atoms exist was an authentic candidate to be interpreted realistically. The reason Perrin didn't argue in this way was not because a belief in atoms lacked the requisite unanimity. The reason is that arguing in this way would not have made much sense, as though deciding empirical issues involves identifying a consensus with the views of other scientists, even if they arrived at their views in a different sort of way. Perrin's task, as we saw, was to settle on an empirical method for calculating Avogadro's number that was eminently reliable and to then use this method to test inferences made on the basis of the atomic hypothesis as applied to diverse physical phenomena. It is because he was successful in doing this, and was the first to do so, that he won the Nobel Prize.

It is true that prior to Perrin there were other estimates of Avogadro's number that were not significantly different from the number at which he arrived. Was preservationism a reason for the satisfactoriness of Perrin's estimate? Even if it was, it is certainly not true today that the satisfactoriness of our current estimates is a product of an analogous argument from preservation since, as we saw, the current estimate of Avogadro's number is more accurate than Perrin's by many orders of magnitude. That is, what is preserved—if any value for Avogadro's number is preserved—is a number far too inaccurate to be of much significance for us today.

Dark matter and dark energy. Both dark matter and dark energy are classic cases where preservationism has little play in scientific progress: Both are entirely unique forms of substance, never thought of, or even conceived of, prior to the 20th century. At least until Fritz Zwicky's postulation of dark matter in 1933, it was usually assumed that the (luminous) matter we interact with on a daily basis is the 'typical' form of matter. In this respect all previous physical theories are wrong, indeed substantially wrong, once we include in the universe's taxonomy dark energy as well as dark matter. Both new entities are manifestly distinct from luminous matter and currently are thought to make up 95% of all the 'stuff' of the universe, as compared to 0% on the previous conception (where 100% of the universe was thought to be luminous). Now it's true that each of these entities has some modicum of theoretical heritage: Dark matter was hypothesized by Fritz Zwicky in the 1930s and has been the subject of some astrophysical interest since then, and dark energy could be said to have made an initial appearance as Einstein's cosmological constant postulated in the 1910s (of course, dark energy may turn out to be an entirely different thing than the cosmological constant). But it is only by means of the telescopic observations we have recounted—observations of the Bullet Cluster (with dark matter) and of high-redshift SN Ia (with dark energy)—that both of them gained anything near a solid reputation. So, with each, we can say that the orthodox view of the composition of the universe was completely displaced and not preserved in the least. Also, with each, we can say that the lack of preservation does not pose a problem for the participant scientists who regard the cases for the reality of these entities to be based purely on novel empirical evidence.

Overall then, with the case studies covered in this book, we have a significant historical argument against the claim that preservationism is a notable feature of scientific advance. The discoveries of mesosomes (or that mesosomes are artifactual), WIMPs, dark matter and dark energy were manifestly not preservative: They each generated a scientific result that was well received by scientists but that also involved a form of doctrinal breach where a prior view of the world was in an important way abandoned for the sake of a new kind of entity not previously anticipated. The case with atoms is a bit different, in that atoms were certainly not uncommonly believed in prior to Perrin's work. Nevertheless, Perrin's justification of their existence was entirely novel, based on an unanticipated analogy between emulsions and (molecular) solutions. In other words, if we take the history of science to guide our philosophical perspective—at least the relatively recent history of science I have examined in this book—it follows that preservationism is a dubious interpretive tool where science makes new and bold advances.

REALISM, THE PESSIMISTIC META-INDUCTION AND PRESERVATIONISM

Of course, the reason preservationism has such wide popularity is because it is thought to provide an effective rejoinder to the pessimistic meta-induction. Thus, if the cases studies we have examined are at all representative of the pattern of scientific progress, then there is the potential here of magnifying the force of the meta-induction. For example, in the case of dark matter and dark energy, astrophysicists had been for a very long time completely mistaken about the nature of the material (both matter and energy) that makes up the universe, having assumed it to be luminous and so having missed up to 95% of it. Their ignorance is even more profound should WIMPs exist—never until now did we even suspect that our galaxy is immersed in a vast WIMP halo. Similarly, for a very long time people were wrongly dubious about the ability to empirically justify the reality of atoms; as a result, competing theories to the atomic theory were viable until as late as the early 20th century. Finally, if mesosomes had turned out to be real, this would have been another case where a previous, generally held, theoretical perspective—that bacteria are organelle-less—would have been exposed as false on empirical grounds. We then have a noticeable pattern with scientific progress that since arriving at a completely novel view of the physical world is often premised on the falsity of a previous, perhaps fundamental theory, it follows that progress is often preceded by substantial ignorance on the topic at hand. As such, it follows pessimistically that these novel advances will themselves likely turn out to be radically false as further discoveries are made, because as we continue to acquire scientific success we correlatively learn more about the failings of past theories and their ontologies. It is thus to be expected that current scientific theories will be found to be false once future, more fundamental progress is made.

But surely this line of reasoning doesn't make sense at all, and the logic of the pessimistic meta-induction, so construed, is giving us the wrong lesson about novel advances. To illustrate, imagine the scientists in the dark matter case reasoning to themselves as follows: 'If we are right about dark matter, then all our predecessors have been mistaken about the nature of physical matter; so we should assume pessimistically that we are mistaken as well'. Surely scientists would view such reasoning as bizarre and excessively abstract. It means that novel advances, by being novel (and so substantially correcting a previous theory), would contain the seeds of their own refutation through the logic of a pessimistic meta-induction. As a result, the safe passage (it is said) to a defensible scientific realism is to not disrupt the epistemic status of past theories (or at least not to disrupt certain chosen components of past theories) but to doggedly 'preserve' the truth of these theories (or at least to preserve the truth of certain chosen components of these theories) to ward off a negative induction. Surely, though, this is a completely wrong-headed view of science: It is a view that counsels scientists to avoid novelty, if such novelty presupposes the substantive falsity of theories that came beforehand; and it is a view that rewards the conservation of theories not because such theories have a particular epistemic value but because their conservation allows scientists to avert a troublesome, philosophically motivated pessimistic meta-induction.

At this stage, the preservationist defender of realism might complain that I am misconstruing what preservationism is trying to do and misrepresenting the task of defending realism (generally speaking) in the philosophy of science. Indeed the job of the philosophy of science is not to counsel scientists on how to do their work or give advice on how they should reason. Doing that, one might suggest, would give an unfamiliar twist to the debate. The usual view is that realism has to do with the interpretation of theories, not with their pursuit, and so my presentation of theoretical preservationism as having prospective aims for future scientific discovery effectively misrepresents the dialectic in the realism debate: The preservationist per se is typically not construed as supplying a positive argument for realism at all but only a response to the antirealist attempt to undermine the realist's no-miracles argument, an argument that notably hinges on the explanation of past scientific success.

Now there is one part of this objection that cannot be denied: Scientists don't construct observational procedures with the goal of preserving past theoretical insights. There would be no need to construct observational procedures if scientists had such a goal, for the result of constructing such procedures would be known from the start: Past theoretical insights will be (found to be) preserved because that is the intellectual design of scientists. Thus, the doctrine of preservationism, as a philosophical thesis, does not translate into a methodological maxim or a tool for discovery that scientists would apply in practice—that is, the philosophical doctrine of preservationism is retrospective, not prospective. But now we have a problem in terms of an understanding of scientific realism, in that realism in the hands of scientists is unremittingly a prospective enterprise: Scientists regard the truth about the world as relayed through the future testimony of innovative observational procedures, not through an intensive theoretical deliberation on past observed results. To this degree, the following assessment of the thesis of scientific realism, as characterized by Ernan McMullin (1984), is misguided:

> Realism is not a regulative principle, and it does not lay down a strategy for scientists.... [Realism] does not look to the future; much more modestly, realism looks to quite specific past historical sequences and asks what best explains them..... The realist seeks an explanation for the regularities he finds in science, just as the scientist seeks an explanation for regularities he finds in the world. (34)

McMullin's assessment is on track if by 'realism' one means 'preservative realism', the sort of realism philosophers are typically (and wrongly, I believe) concerned with. It is true that preservative realism is not regulative for scientists: Scientists don't strive to preserve theoretical insights but rather keep their minds open in the context of a contingent empirical inquiry. Moreover, it's true that preservative realism aims to be empirically based, for the no-miracles argument for realism that underlies preservation is itself a form of empirical argument. Given that a scientific theory has been successful in the past (an empirical claim), and given that the best explanation for this contingent success is that scientists have in some way latched onto the truth, it follows that we have support for the truth of this theory. However, if we abandon preservationism and adopt a prospective realism, then the philosophic task of retroductively explaining past scientific practice—the task of McMullin's realist—becomes pointless, just as it is pointless for a scientist to be exclusively preoccupied with the interpretation of past, empirical regularities. Rather, the greater preoccupation of scientists is to construct novel observational procedures that generate new and informative empirical information. Should they happen to reflect on past observational results and retroductively explain them in theoretical terms, that will only serve as a precursor to further observational interventions. Accordingly, it is incumbent upon the philosopher who defends a (prospective) realism to examine what scientists are currently doing and not dwell on theoretical claims that have been preserved throughout the history of science, since the justifiedness of a theoretical claim for a scientist is not based in its historical persistence or on what used to be regarded as its empirical support but is instead based on what counts as its current empirical support.

There is a related confusion in which McMullin and other preservative realists engage. Once again, on their view, scientific realism stands as an empirical thesis, one that can be confirmed or falsified by an examination of scientific practice. McMullin (1984) comments:

> What we have learned is that retroductive inference works in the world we have and with the senses we have for investigating that world. This is a contingent fact, as far as I can see. This is why realism as I have defined it is in part an empirical thesis. There could well be a universe in which observable regularities would not be explainable in terms of hidden structures, that is, a world in which retroduction would not work.... Scientific realism is not a logical doctrine about the implications of successful retroductive inference. Nor is it a metaphysical claim about how any world must be.... It is a quite limited claim that purports to explain why certain ways of proceeding in science have worked out as well as they (contingently) have. (29–30)

What McMullin is suggesting is that a preservative, or for him structural, realism is not a sure conclusion that results from a reflection on scientific advance. Surely this is true. It may turn out that the theoretical structures McMullin claims we retrospectively find, for example, in the geologic time-scale, in the structure of cells and molecules and so on are repudiated with subsequent scientific advances, leaving even the preservative (and structural) realist to concede the power of the pessimistic induction. But this is a concession we would be forced to take only if we are preservative (or structural) realists. In other words, I don't see any of the scientists we discussed in our episodes recoiling from realism when they encounter substantive theoretical change, nor is there any substantive reason why they should recoil, given that the preservation of prior theoretical conceptions is not a priority. In this regard, consider again our historical episodes and the questions with which scientists are faced. Are mesosomes real? Are WIMPs real, and are they really interacting with a detector? Are atoms real? Is dark matter real? Is dark energy, or at least the accelerative expansion of the universe, real? These are the core questions with which the scientists are fundamentally preoccupied, and should any of these questions be answered in the negative, there was never the option for any of these scientists to say that there isn't any reality after all, or that science ultimately doesn't have the capability to correctly describe reality. Reality simply turned out to be somewhat different from what was expected. From here, scientists simply return to their labs, telescopes and so on and devise new ways of exploring the physical world.

In light of the fact that a prospective realism may be better suited as an interpretation of scientific practice, why would philosophers such as McMullin and so many others cleave to a theoretical preservationism? My suspicion is that they are led to advocate preservative realism because they are attracted to a form of robustness reasoning at the theoretical level. That scientists from differing time periods; in different social, professional or cultural contexts; and using varied experimental and mathematical apparatus arrive at the same preserved, theoretical conception shows that this conception is 'robust', and for preservationists this means that these scientists are catching on to something real. To adapt a phrase used by the physicist David Cline when, on the basis of his experimental work, he came to reluctantly accept the reality of neutral currents, '[We] don't see how to make these effects go away' (Galison 1987, 235). It's that sort of thinking robustness theorists and their preservationist kin find compelling. If observed results keep coming back despite a change in observational procedure, or if certain theoretical conceptions keep reappearing even with a change in scientific tradition, it follows (it is claimed) that what we're observing or conceiving of is real. Of course, the main burden of this book has been to dispel this logic as regards observational procedures—and the logic is no better at the theoretical level. Alternatively, on my view, the mindset of an observational (or experimental) scientist is such that she is not at all averse to discarding past observed results or past theoretical conceptions, if the empirical facts so dictate, and she will compensate by becoming a realist about newly acquired, contrary results or conceptions. Indeed, it can sometimes happen that the path to an accepted scientific result is highly unique, due to its heavy dependence on a particular observational procedure or theoretical conception. As such, it is all too easy to 'make the [result] go away' by not strictly adhering to a certain observational or theoretical protocol. My point is that this procedural dependence is no bar to being a realist about the generated result (neither for a scientist nor for a philosopher). Whether one should be a realist about the result depends on the details of the protocol at hand and its current level of empirical support, not on the ability of the result to make a reappearance under differing circumstances.

Given the intensive preoccupation many philosophers of science seem to have with preservationist (and structural) forms of realism, it is no wonder that the musings of philosophers on issues of scientific realism are routinely ignored by scientists. This is not to deny that, in certain specialty areas, such as evolutionary theory or quantum mechanics, a great deal of disciplinary overlap can be found that equally engages both philosophers and scientists. But, by and large, contemporary scientific research proceeds in complete ignorance of philosophical ruminations concerning scientific activity. For me this is a concern when one thinks of the social value of philosophy of science. But then one might suggest, in response, that philosophers shouldn't concern themselves with what scientists are currently doing and that instead their topic is retrospective, looking back and logically reconstructing scientific work in an internal history or providing a social narrative on past scientific practice in an external history. If this is the response of philosophers, then they shouldn't be surprised when present-day scientists express amazement that philosophers consider themselves to be studying science at all in their avoidance of actual, recent scientific practice. Science, scientists believe, has progressed to such a degree that old science is often not even recognizable as science any longer, such as with caloric or ether theories. By comparison, many philosophers are still taken by the pessimistic meta-induction with its focus on past theories that have long since disappeared from the scientific scene. If one thinks that the theories of caloric and ether are representative of 'quality' science, then one might well be impressed by how successful science can be false and nonreferring. Of course, in no sense am I claiming that scientists have arrived at the 'absolute truth'. Every scientist knows that science is fallible and that future progress may reveal that our current theories and discoveries are mistaken, just as we now think of caloric and ether theories as mistaken. In fact, this is the lesson we should take from studying the history of science with its host of refuted entities—we should always be prepared to learn that the current scientific orthodoxy is false. For its part, theoretical preservation, where certain claims concerning the existence of theoretical entities persistently hold true as science progresses, just doesn't obtain very often in scientific research, especially when we are dealing with fundamental scientific discoveries. Of particular note here is what we discovered in our survey of recent developments in astrophysics: The accepted understanding of the ultimate taxonomy of the universe has surprisingly shifted from asserting that 100% of all matter is luminous to claiming instead that 5% of all matter is luminous, with novel forms of dark matter and energy filling the 95% gap. Scientists, let me repeat, are not dissuaded by such radical conceptual change and feel no urge to be realist about (structural) components of past theories in an effort to explain past 'successes'. This is because they are ruthlessly forward-looking in their realism, and not backwards-looking, as preservationist philosophers tend to be.

Nevertheless, dispensing with the pessimistic meta-induction is not quite that easy, and we are still left with the lingering question of how scientists and their philosophical allies can respond to this challenge. How can one be sure about a realist interpretation of a current theory or of a current observed result, if we concede the extensive history of failure that one finds in the history of science? We should not be hubristic and blandly say, 'Before we were wrong, but now we're getting it right!' That is more an expression of conviction than a philosophical position. Given a past pattern of failed but otherwise successful scientific theories, and given that we have dispensed with the (theoretical) preservationist option, by what entitlement can we be realists about scientific theories? What is the future for scientific realism, if it is without a past?

THE IMPROVED STANDARDS RESPONSE: 'METHODOLOGICAL PRESERVATIONISM'

The answer to this problem is to focus on another kind of preservationism with which scientists involve themselves, which I call 'methodological' or 'standards' preservationism. My assertion is that this form of preservationism has the resources for one to effectively defend scientific realism. My inspiration for this approach derives from Doppelt (2007). Doppelt is similarly skeptical about preservative realism and wants to suggest that subsequent theories may simply be better than their predecessors when judged on the basis of his set of 'standards of explanatory and predictive success' (109), which include the familiar items, 'unification, consilience, simplicity, empirical adequacy, completeness, internal coherence, and intuitive plausibility' (111). Admittedly Doppelt doesn't call his approach 'preservative', though clearly it is in the sense that his chosen set of standards is for him nonnegotiable. What is negotiable for him is how well a science can meet these (preserved) standards. In what he calls the process of 'cognitive progress', there is an 'elevation' of his chosen standards in the sense that prior scientific theories, despite their successes, are not deemed as successful as subsequent, scientific theories since the prior theories are unable to meet the heightened standards set and satisfied by the later theories. For him, this is how one stops the slide to antirealism envisaged by the pessimistic inductivist: One need not worry about past theories that were successful but were false, for these past theories may not be successful after all when looked at from the perspective of the heightened standards of success that current theories meet. Doppelt summarizes his view this way:

> For my brand of realism, the most striking thing about our best current theories is not mere success, or even the fact of more success than predecessors. It is rather the fact that they succeed in both raising and, to an impressive degree, meeting standards of accuracy, unification, consilience, explanatory breadth, completeness, and so forth that are qualitatively far more demanding than all their predecessors either aimed at or attained. (112)

The problem for Doppelt's approach, of which he is aware, is that the improvement of standards has no natural endpoint. One can easily imagine subsequent scientific theories raising and meeting new standards that surpass the heightened standards of our current theories, which leaves us with a renewed pessimistic induction, one that compels us to question the success of our current theories. This new meta-induction, Doppelt notes, is different from the original one and needs to be considered carefully: 'Arguably', he says, 'in the history of the sciences, there is greater continuity in standards of empirical success than in the theories taken to realize them' (113). This is, in my mind, the key to the matter: If there is a continuity of standards, a form of methodological preservation, then there is a way to turn back the pessimistic meta-induction and revive scientific realism without necessarily engaging in a form of theoretical preservation.

However, Doppelt declines to go this route. He chooses to take the harder path, allowing the possibility of a continually ascending set of standards but asserting nevertheless that scientific realism is a reasonable doctrine:

> If and when higher standards and more successful theories appear, this development defeats not scientific realism but rather which theories it is reasonable for the realist to take as approximately true. (114)

He calls this an optimistic perspective, and it sounds commonsensical to the scientific mind. The view is: We are to be scientific realists, but we haven't yet decided what theory to be realist about; that will need to wait until the end of science, when we've arrived at the best theory meeting the best standards. The problem is that scientific realism can't wait that long. The doctrine is asking us to be realist about our current theories, not accept a promissory note whose fulfillment will perpetually occur at an indefinite time in the future, if it occurs at all. Such a note amounts to a realism that says, 'Be a realist about the best theory possible', where 'best' means meeting the best standards possible. This is a form of realism—but not a useful one because of its unreachable exclusivity. It is a form of realism that asks us to be agnostic about the reality of what is described in our current, best theories—yet scientific realism, as usually construed, asks us to be realists about what is described in our current, best theories, to at least an approximate degree.

A further questionable aspect of Doppelt's approach concerns his chosen set of 'standards of explanatory and predictive success' (109). Some of the items have no sure status as preferred criteria for the worth of a scientific theory. For example, simplicity is a notably troublesome standard considering the intellectual complexity of many theories; unification assumes the unifiedness of nature, and there is room to assert in a coherent fashion the irreducibility of some sciences to other sciences; and intuitive plausibility is arguably very subjective, often dependent on one's cultural background. So from Doppelt we at least need to hear more about why he thinks this set of standards is special (and forms the basis for future improvements of standards) and what reasons he gives for thinking that scientists would actually advocate such a list.

Despite the flaws in Doppelt's own approach to standards, it is nevertheless my view that he is on the right track in his approach to defending realism. By focusing on methodological standards and not on the ontological claims themselves (such as 'Is there caloric?' or 'Is there ether?'), he provides a framework that gives us a way to avert the trap set by the pessimistic meta-induction. Where he fails is in focusing too intently in his framework on theorizing and on defining the best standards for a theory. Scientific work often does not involve theorizing or even speculating on the theoretical relevance of acquired empirical data but is instead preoccupied with the improvement of observational procedures. It is often not through theorizing that scientists become convinced of the reality of hypothesized objects but rather through a form of controlled observation. So if the issue is scientific realism, the proper place to look for its legitimacy, most of the time, is in the observational procedures scientists use to confirm or disconfirm the existence of an entity, not in the theorizing that sets the stage for these procedures. Thus, if our focus is on standards, as I think Doppelt is right to suggest, then I think we should look at the standards scientists use in manufacturing observational procedures. It is in the context of these procedures that scientists argue for the existence of theoretical entities, and these arguments manifestly involve standards that share very little in common with the list of standards advocated by Doppelt.

To understand the types of standards in play in scientific observation, I focus again on our case studies. In reflecting on these studies, one very general requirement on observational procedures becomes apparent, one that rests on guaranteed (though perhaps overly obvious) philosophical grounds: Observational procedures must involve reliable processes, processes that at a minimum tend to generate truthful claims. Using such procedures, scientists can engage in what I call 'reliable process reasoning', reasoning that has (as I noted in chapter 2) the rough form:

1. A certain observational procedure is reliable.
2. Using this procedure a particular observational report is generated;

Thus, this observational report is true.

As with the research regarding mesosomes, the relevant report could be one that expresses an existence claim, such as 'some (kind of) object exists' or 'some (kind of) object does not exist'. Such reasoning is ubiquitous in the case studies we have been examining. In each case instruments are designed under the assumption that they will allow scientists to reliably observe theorized objects. For instance, in the search for WIMPs, astroparticle physicists set up experimental detectors deep in mines on the basis of the (reasonable) theoretical assumption that the materials composing these detectors have the capacity to interact with incoming WIMPs; they are thus using a reliable process in the sense that if, alternatively, they put these detectors on the surface of the earth, these detectors would be triggered by many more things than WIMPs. In a similar way, Perrin argues that vertical distribution experiments using emulsions of gamboge constitute an analog to vertical distributions of gaseous solutions and so can play a role in a reliable procedure for determining Avogadro's number; by contrast, using an emulsion that does not exhibit this analogous behavior would have no such value. With mesosomes, determining their reality (or non-reality) clearly requires some way of magnifying the contents of cells—magnification, here, is a reliable process (generally speaking). Finally, with dark matter and dark energy, using light (and other) telescopic observations has an obvious informative value. As we saw, the value of telescopy culminated in the ability to observe the fortuitous cosmic separation of dark matter from luminous matter, as was found with the Bullet Cluster; telescopy also allowed astrophysicists to detect the accelerative expansion of the universe, facilitated by measuring the luminosity of distant supernovae and comparing these measurements with what is expected on various models of universal expansion. All of these cases involve in this fundamental way a form of reliable process reasoning, though of course the full details of this reasoning in each case is much more elaborate, and to be sure the relevant form of reliability is ultimately comparative (e.g., examining cellular structure using magnification versus without magnification, detecting WIMPs in a mine versus on the earth's surface and so on).

The first obvious point to make here is that reliable process reasoning, considered abstractly (as in my schema), is the core of any scientific observational procedure and is thus a preserved methodological requirement. But that isn't, really, saying very much. Even the proponent of robustness is an advocate of reliable process reasoning—either he assumes the minimal reliability of observational strategies that converge on a single result or the convergence of these results is how he demonstrates the reliability of these strategies. The more suggestive point that I wish to make here is that the particular instantiations of reliable process reasoning as we described them in the previous paragraph relating to our case studies are also preserved over time. For example, so long as WIMP detectors are unable to distinguish between WIMPs and other cosmic particles such as muons that can cause false WIMP detection events, it will always be a more reliable approach to put detectors deep in mines that shield WIMP detectors from cosmic particles. Also, when considering Perrin's vertical distribution experiments using gamboge emulsions, drawing an analogy between vertical distributions of gamboge emulsions and similar distributions of gaseous solutions (so long as the analogy holds up) will always be a reliable approach to determining Avogadro's number, keeping in mind the quantitative limitations of such an approach. Similarly, no one would ever doubt that magnifying the contents of a cell, as a general strategy, is a reliable approach to ascertaining the nature of these contents, nor would anyone doubt that telescopic observations form part of a reliable procedure for studying the potential reality of both dark matter and dark energy. In all of these cases, a methodology is established that, from the perspective of reliable process reasoning, has an assured (though admittedly limited) ability to reveal actual states of the world, an ability that we expect to last into perpetuity (assuming that native human observational functionality does not itself change over time).

To give a sense of the importance of such core methodologies, consider the process of 'naked-eye' or 'unenhanced' (i.e., to include other modalities than vision) observation. This is our first and most important observational method, considered to be reliable for as long as anyone can remember and still reliable to this day. Moreover, no one is ever going to fundamentally subvert the reliability of naked-eye observation as it forms the empirical basis for all our interactions with the world. If we were to deny the reliability of naked-eye observation (at least tacitly), we would lose all epistemological bearings with respect to the world. Its basic and continued status as reliable is so assured that there is a form of philosophical theorizing, called empiricism, that views naked-eye observation as the only source of reliable information. The case of naked-eye observation is instructive because, despite its reliability, there are plenty of cases one can cite in which this reliability is suspect. Descartes' *Meditations* contains the classic expression of this sort of worry—as the first meditation suggests, there are too many instances where sensations and perceptions have fooled us for us to feel much comfort about their reliability as sources of information. Scientific progress itself has undermined the reliability of observation, announcing that the various secondary qualities that enrich our sensory lives are illusory and that the physical world is actually quite colorless, odorless and tasteless. But these facts have done nothing to shake our confidence in naked-eye observation, and scientific, empirical research is almost paradoxical by denying on the one hand the reliability of what is observed (in affirming the reality of the 'scientific image') and on the other hand relying absolutely on the reliability of what is observed (in its methodological dependence on empirical facts).

I believe a similar assessment is applicable to the other sorts of reliable processes described above relating to our case studies. Each of them, though fundamentally reliable, is subject to correction. Magnification in the mesosome case moved from the light-microscopic to the electron-microscopic, a clear boost in performance when examining cellular substructure. The preparative methods needed for electron-microscopic investigation also changed over time. Specifically, the use of the R–K method with its dependence on OsO_2 fixation came to be regarded as generating artifacts and was subsequently replaced with freeze-fracturing and freeze-substitution methods that were considered more reliable. In WIMP research, once the detectors were placed in mines, various improvements occurred, such as correcting for PMT noise (UKDM), rejecting surface electron events (CDMS), installing a scintillating muon veto (EDELWEISS) and carefully excluding ambient radon (DAMA). In Perrin's use of a vertical distribution experiment, his choice of what ideal size of gamboge grain to use was subject to correction, going from .212 microns (Perrin 1910) to .367 microns (Perrin 1916). Regarding the existence of dark matter, telescopic observations that confirm the existence of dark matter were provided an enhanced reliability once the Bullet Cluster was discovered that exhibited the separation of dark matter from luminous matter. Finally, telescopic observations revealing the accelerative expansion of the universe were made more reliable when proper accounting was made for a variety of possible sources of error, notably the presence of cosmic dust and cosmic evolution. What we find in all these cases is that the reliability of an observational process is corrected or enhanced without the reliability of the underlying observational process being questioned. The status of the underlying observational methodology as reliable is 'preserved', we say, despite these corrections or enhancements. I call such an approach 'methodological preservationism'.

We might note here that the cases of dark energy and dark matter, as we have described them, involve an observational strategy of a more abstract kind, which we called targeted testing. In both cases there are competing descriptions of a set of observed results: The rotation curves of spiral galaxies and the velocity distributions of galaxy clusters, for example, could be viewed as manifestations of dark matter or of a modifiable force of gravity, and when observing high-redshift SN 1a, the dimness of these supernovae could be a product of an accelerative expanding universe or of cosmic dust and evolution. With targeted testing, an observational procedure is employed that resolves this underdetermination: It is a procedure that, to begin with, does not contest the truth of each of the competing hypotheses but that is able to marshal empirical support for one of the competing hypotheses (as opposed to the other) nevertheless. A form of (abstract) reasoning similar to targeted testing can be found in the WIMP research we have examined. Recall that DAMA is concerned about the large variety of model assumptions astrophysics groups need to make to identify individual WIMP events. So many assumptions are needed, and there is so little knowledge about when exactly these assumptions hold, that DAMA is skeptical about reaching any definite conclusions about the existence (or not) of WIMPs on that basis. Many of these model assumptions deal with possible sources of error affecting the performance of the detectors placed at the bottom of mines, errors caused by such things as ambient neutrons, electromagnetic influences, radon gas, muon flux and so on, and both DAMA and the competing WIMP search groups do their best to minimize these influences. DAMA's ingenuity in this regard involves recording possible WIMP detections over the course of years and seeing if these detections exhibit a pattern representative of both its understanding of the WIMP halo (i.e., as theoretically enveloping our galaxy and solar system) as well as its knowledge of how this halo potentially interacts with Earth-based detectors, that is, as leading to an annual modulation in possible WIMP detection events. In finding this annual modulation, DAMA's conjecture is that the cause of the modulation is the presence of incoming WIMPs, since any item on its list of possible errors, should it occur, would lead to a different sort of modulation, if any modulation at all. That is, even if we suppose that these sources of error are in play, the witnessed modulation still indicates the presence of incoming WIMPs; for this reason, DAMA describes its annual modulation result as model independent. By contrast, the strategies used by the competing research groups (UKDM, CDMS and EDELWEISS) are model dependent in that their results are heavily dependent on the particular model assumptions they need to make, particularly concerning the presence or absence of potential sources of error. So long as DAMA is right in its understanding of the source of the modulation as well as about how its detectors work, the value of its work in the context of future observational work regarding WIMPs is assured, even if it is the case that it is eventually improved upon by model-dependent approaches that succeed in specifically identifying individual WIMP detection events, or even if it comes to pass that there are no WIMP detection events after all. In the latter sort of case, DAMA's work is significant enough that its errors would need substantive 'explaining away'.

Another abstract methodological tool that can be used to enhance the reliability of an observational procedure involves 'calibration', a strategy that (I argue) Perrin utilized in arguing for atomic theory. For example, Perrin verified the accuracy of Einstein's equation (E) by showing that it generates the same result for Avogadro's number (*N*) as that obtained by Perrin's vertical distribution experiment—that is, we calibrate Einstein's method for determining *N* by exhibiting its consistency with another approach whose reliability is not subject to scrutiny. Generally speaking, calibration has a host of applications whereby the observed results of an observational procedure of uncertain reliability are given an enhanced confirmation by showing that the procedure generates other sorts of observed results whose accuracy is confirmed through the application of a more reliable (calibrating) procedure.

Overall I have been arguing that the case studies presented in this book each exhibit a significant methodological, observational advance; new observational procedures are established with a level of reliability that can be anticipated to persist into the future. The classic example of a highly preserved and informative (albeit primeval) methodology is naked-eye observation, whose reliability no one rejects (despite its celebrated flaws). Close relatives to unenhanced observation involve the use of magnifying devices (microscopes) for investigating cellular substructure (and other microscopic phenomena) and telescopes for astronomical observations. More detailed observational advances include the use of freeze-fracturing and freeze-substitution for the preparation of bacterial specimens in order to ascertain whether they contain mesosomes, Perrin's use of gamboge emulsions as physical structures analogous to molecular solutions with the goal of calculating Avogadro's number and the use of detectors located deep in mines for the purposes of observing distinct kinds of cosmic particles such as WIMPs (as opposed to other 'undesirable' cosmic particles, such as muons, which are largely intercepted by the intervening rock). We also cited more abstract, reason-based methodological tools, from reliable process reasoning to targeted testing and calibration. In the course of an empirical investigation, one always has the option to return to these methods (if they can be applied) to gather at least minimally reliable information about a designated subject area. In this respect these methods are preserved. Moreover, this preservation of methods may not correspond to a form of theoretical preservation. Taking again the base case of naked-eye observation, such a procedure is always considered a source of reliable information, even though over the course of history many different theories have arisen that give conceptual substance to what it is that our naked-eye observation reveals to us. Depending on the theory of the time, our scientific conception of what it is we are observing can change from being composed of atoms to not being so composed, from containing caloric to not containing caloric, from requiring the presence of a luminiferous ether to not requiring ether and so on. In other words, the preservation of observational methods that I assert is integral to scientific research does not necessarily correspond to the preservation of certain theoretical entities (or structures), as is required by the usual preservationist defenders of scientific realism.

That is not to say that the observational procedures that have been preserved have no ontological significance. Quite the contrary: Objects revealed by such preserved methods acquire a prima facie claim to reality that counterweighs the negative historical induction that would lead one to assert their nonreality. This is exactly the case with naked-eye observation, where the drastic changes in scientific theory highlighted by the pessimistic induction fail to subvert in our minds the reality of the objects we observe with our bare modalities. For example, we continue to observe the thoroughgoing solidity of chairs and tables, despite an atomic theory that tells us to regard such objects as mostly empty space. We continue to observe and assert the objective reality of colors and smells, despite holding to psychological theories that place such qualities subjectively in the mind. Similarly, telescopic observations reveal the presence of distant stars and galaxies, and we feel confident in the existence of such things, even if astronomical theory tells us that these cosmic entities no longer exist or are at least drastically different from how we see them. Preserved methods can even recommend contrary ontologies, with each ontology holding a realist sway on our minds. One of the classic cases of such an occurrence is the cellular structure of our skin: Our skin to the naked eye appears to be a simple thin sheet, and we express initial surprise to learn that it is composed of innumerable distinct cells as revealed by magnification.

Nevertheless, we still feel comfortable about referring to and seeing skin as a thin, unsegmented sheet (we 'peel it off as a homogenous layer' when skinning animals, for example), a realist attitude that persists despite our scientific enlightenment. This is because the preserved method of naked-eye observation has the epistemic authority to present to us an ontology that has a claim to be taken realistically. The quality of this method (or of any preserved method) inclines us to be prima facie realists about the entities that the method presents to us, and where the testimonies of our preserved methods conflict, we can be prima facie realists about conflicting ontologies.

However, there is no denying the possibility that the objects revealed by purportedly preserved observational procedures could in due course be shown to be unreal and the procedures themselves cast aside. This is what happened with the 'discovery' of mesosomes using the R–K method. The R–K method was for a significant period of time the standard method of preparing microbiological specimens, and arguably a belief in the existence of mesosomes was initially commonplace, since they were witnessed by means of the R–K method. Thus, opinions of microbiologists during the 1960s would have been that the R–K method is a preserved observational method whose reliability has an assurance to continue into the future. Despite this opinion, we know that such scientists were wrong. The authority of the R–K method came under dispute regarding the question of bacterial substructure, with the result that mesosomes were written off as artifacts once the R–K method fell into disuse (at least regarding the study of bacterial mesosomes). One can identify a similar turn of events with the experiments performed by Victor Henri that predated (by a few years) Perrin's own experiments on emulsions. For many physicists at that time, Henri's experiments constituted an established methodology with definitive observational implications, and these experiments led many physicists to question Einstein's theoretical interpretation of the diffusion of emulsive grains (as set forth in Einstein's equation [E]). Yet the authoritative status of Henri's experiments was short lived, lasting only until Perrin's work both exposed errors in these experiments and, for its own part, vindicated the accuracy of Einstein's equation (E).

Accordingly, in light of the possibility that even (relatively) preserved methods might be disputed, one might feel pressured to argue as follows.

Since so many presumed methodological, observational advances have been found to be flawed (and here one might cite the various cases where even naked-eye observation goes wrong), there is no assurance that what we currently count as such an advance will itself be safe from refutation. Hence, by an analogous argument to the pessimistic meta-induction, there should be no anticipation that any observational method is perpetually preserved—and from here one might argue that the reality of objects as revealed by observational procedures is itself thrown into doubt, since, from one procedure to the next, we find ourselves committed to different ontologies. In turn, without the proposed preservation of methodologies as I have suggested, and so without being assured of the prima facie reality of the objects revealed by preserved methodologies, we are no further along in resolving the pessimistic meta-induction on behalf of realism.

Now there is reason to resist such a skeptical conclusion about the value of methodological preservationism, for many observational methods have turned out to be quite difficult to dispense with. Naked-eye observation, for one, will never be cast aside despite the many instances in which it has been shown to be illusory, nor will anyone suggest that we should stop magnifying microscopic entities or stop using telescopes to examine celestial objects, despite the errors to which these procedures are prone. Alternatively, some prior observational procedure may be dispensed with, but not because of its intrinsic unreliability; instead it is replaced by a more precise method. This may yet occur with the search for WIMPs. DAMA defends the use of its model-independent, annual modulation approach to identifying incoming WIMPs, even though it would surely agree that a model-dependent approach would be preferable as an observational proof of the existence of WIMPs (despite its reliance on a plethora of controversial model assumptions) so long as the reliability of this approach could be assured. Similarly, Perrin's estimate of Avogadro's number by means of his emulsion experiments is no longer the method of choice for modern scientists determining this number. But the problem with Perrin's approach is not that it is wrong—it is simply too imprecise for modern scientists who can now calculate Avogadro's number to many more decimal places. Again, consider the strategies used by astrophysicists to account for the influence of dust and cosmic evolution on observations of distant supernovae. First, with SN

1a observations there is a need to account for and (potentially) correct the effects of dust and evolution. When HZT succeeded in observing extremely high redshift SN Ia in the early 2000s, noting that these SN Ia were brighter than expected, they succeeded in thereby discounting the effects of dust and evolution. This was a substantive, methodological accomplishment: HZT's high redshift observations set forth a strong rebuttal of the dust and evolution alternatives (assuming of course that their procedures were carried out correctly), and we can anticipate that future astrophysicists will acknowledge this fact. But this is not to deny that subsequent procedures to measure the effects of dust or evolution might be much more precise than HZT's, or that HZT's results might be interpreted in a different way (e.g., as refuting rather than supporting the existence of dark energy). These sorts of happenings are compatible with the substance of HZT's accomplishment—the design of an observational procedure capable of revealing that extremely high redshift SN 1a are brighter than anticipated, an accomplishment whose merits have a lasting value for astrophysicists. The case of dark matter is similar. With Clowe and colleagues' observations of the Bullet Cluster, one can clearly discern the disjoint locations of a galaxy cluster's luminous matter and the bulk of its gravitational potential. In effect, their strategy to utilize the Bullet Cluster in settling the question of whether there is some kind of matter other than luminous matter achieved a key and lasting methodological advance whose merits are assured, despite the fact that there are still interpretative issues to be settled (Moti Milgrom [2008], we recall, embraced the result as proving the existence of dark matter but maintained that the gravity law might need adjustment as well), and despite the fact that their methods can be improved upon by future work.

What I am suggesting is that whereas past scientific change has often involved a complete change in the scientific catalogue of 'real' objects, with methodologies there is correspondingly less change and more continuity. Here I am drawing inspiration from Doppelt's (2007) insight that there is a difference between the sort of pessimistic induction that applies to observational standards and the original one dealing with ontologies. Doppelt's suggestion is that there is 'greater continuity in standards of empirical success than in the theories taken to realize them' (113), a claim I am supportive of as regards certain preserved observational strategies: Naked-eye observation, as the base case, has always been and always will be a core observational strategy, even as scientists demur on the reality of the observed properties revealed in this fashion (e.g., even though it is sometimes denied that observed colors, tastes, sounds, feels and smells are 'real', no one proposes to dispense with the faculties that generate these qualities). But despite that 'greater continuity', we must still allow the possibility that any particular observational standard can be displaced, a possibility Doppelt (2007) acknowledges. He comments:

> For the sake of the argument, imagine that there is inductive evidence that the high standards met to a significant degree by our best current theories will in all likelihood be superseded by yet higher standards that our best current theories will be unable to meet and that yet new theories will be able to meet. (113)

For Doppelt, this renewed challenge to realism is met by being a realist about whichever new theory meets the new higher standards:

> If and when higher standards and more successful theories appear, this development defeats not scientific realism but rather which theories it is reasonable for the realist to take as approximately true. (114)

In other words, as observational (and theoretical) standards change, what theories we should be realist about also changes, whereas the thesis of realism itself still stands. In my opinion, to enact such a (perhaps very natural) rejoinder to the pessimistic induction, one doesn't actually need to make recourse to the question of observational standards, for one can always demur on what theory one should be realist about. However, in the end, such a response to the pessimistic induction doesn't carry much philosophical weight, since realism, as usually understood (i.e., not in the way van Fraassen (1980) understands realism, as simply holding that the aim of science is truth without claiming that science has actually arrived at the truth), doesn't suggest that the best theory possible (constructed sometime in the future) is true or approximately true but rather that this is the case about our current (best) scientific theories. As such, a realist that permits the possible, absolute falsity of our current, best theory surely concedes too much: If we have no idea about what theory we should be realist about, we may as well be nonrealists. In other words, the doctrine of realism must amount to more than just the issuance of a promissory note concerning the reality of some future-conceived objects.

Alternatively, my approach to resolving the pessimistic induction is different from Doppelt's (2007). Whether one advocates a pessimistic induction that depicts radical change in our observational standards or the original pessimistic induction that depicts radical change in our ontologies, one notes in either case the empirical success of many sciences (which realists claim is best explained by the theories being true) and then highlights the fact that the history of science contains many examples of empirically successful theories that have turned out (in retrospect) to be false. Preservationists (of the ontological sort) then respond to this situation by locating parts of theories that have never been falsified, thus breaking the induction at that point; the success of these parts (in the context of the larger theories to which they belong) is thus (presumably) never detached from their truth. Of course, as I have suggested, such recourse to the preserved parts of theories is a hazardous strategy, as there are many reasons why a part of a theory may be preserved, none of which have to do with the part's truthfulness (as we saw Stanford and Chang arguing above; see also Votsis 2011, 1228–1229). Again, a robustness-type argument at the theoretical level is no better than such an argument at the observational level: Arguing from the preservation of theoretical commitments to a realist interpretation of these commitments is uncertain given the plethora of cultural factors that influence the choice of theories. Surely, a better indicator of the truthfulness of a preserved part of a theory would be to describe how the inclusion of such a part serves to enhance this theory's 'explanatory and predictive success' (adopting for convenience Doppelt's favoured criteria) rather than simply pointing out that this part has persisted from previous theorizing.

It is precisely here that methodological preservationism can make a contribution to resolving the problem posed against realism by the pessimistic induction. We note, to begin, that certain observational procedures are preserved (to varying degrees) across a variety of sciences; some have become so common that they constitute 'standards'—observational methods that foreseeably will never be usurped. Still, we acknowledge the fallibility of any observational standard. For example, we acknowledge that microscopes constitute a reliable observational procedure in investigating microscopic reality, even though microscopes often provide misleading information. The characteristic aspect of observed results, though, as compared to theoretical claims, is that they cannot be said to be empirically correct if they turn out to be fundamentally mistaken. For example, if it turns out that microscopes completely mislead regarding the nature of microscopic reality, then it doesn't make sense to say that microscopes constitute an empirically successful observational procedure. Now recall the logic of the pessimistic induction: The past contains a vast litany of empirically successful though false scientific theories, and so it is highly unlikely that future empirically successful theories will turn out to be true. The analogous, pessimistic inductive argument dealing with observational procedures is accordingly this: The past contains a vast litany of empirically successful observational procedures that generate (nevertheless) false observed results, and so it is highly unlikely that future empirically successful observational procedures will also generate true observed results. It should now be clear that such an analogous argument completely fails. The premise is false, because the past cannot contain empirically successful observational procedures that generate false observed results—simply, the falsity of the observed results of an observational procedure implies the failure of this procedure. Moreover, the conclusion is false for the same reason: By the very meaning of empirical success, future empirically successful observational procedures must generate true observed results (at least most of the time, as success is compatible with occasional failure). Simply put, the pessimistic induction does not have an application when we move from (empirically successful though false) theories to (empirically successful though falsity-generating) observational methods.

Recognizing the irrelevance of the pessimistic induction (as applied analogously to observational methods) to methodological preservationism is paramount. I am not suggesting that some empirically successful observational procedures need to be preserved in order to be assured of the (preserved) truth of certain observed results. Rather, the motivation for methodological preservationism is the recognition that some, often very familiar observational methods have achieved the status of being iconically reliable, in the sense that they constitute a resource that is perpetually available for the purposes of generating information about the world. They are, in this regard, our first guide to the nature of reality and sometimes even provide an ontological perspective that is in general terms persistent and stable.

Nevertheless, could it not still happen, as per the skeptical instinct underlying the original pessimistic induction, that even our most highly preserved observational procedures are radically mistaken and habitually generate false reports about the world? Even if we accept that an empirically successful observational procedure cannot (in general) generate false observation reports, how can we be sure that our favoured set of preserved observational methods is not empirically unsuccessful after all and indeed systematically generates false observational results?

At this stage, we arrive at the cusp of a severe, systematic skeptical view of the world that strains at credulity. Could naked-eye observation, in its larger aspects, be fundamentally mistaken? Could astronomical observation become less certain by the use of telescopes? Are microscopes systematically misleading us about the nature of cellular reality? These are possibilities, but they need not be taken seriously if we are to make the first step in understanding science. By comparison, in understanding science, there is no comparable first step in establishing a preserved ontology. Scientists do not strain at credulity in suggesting that we have made broad errors in understanding the ontology of the world. It may be that atomic theory is false, that there is no dark matter or dark energy and that in fact we are but dreamy characters in the Creator's mind. None of this is of concern to the scientist who simply seeks the truth about the world, whatever this truth might be. For the scientist, there is no first, preserved ontology to which we must be committed, not even a structuralist one. Rather, there is a first, preserved methodology—the methodology of naked-eye observation—and on this basis a series of further, fairly uncontroversial preserved methodologies that involve either reason-based enhancements to (such as reliable process reasoning, targeted testing and calibration) or technological modifications of (such as telescopes and microscopes) the original method of unenhanced observation.

CONCLUSION

The main aim of this book has been to cast doubt on the purported epistemic value of robustness reasoning. To be clear, I do not deny that there could be epistemic value in utilizing different procedures to generate an observational claim—for example, one procedure might be used to calibrate or target test another procedure. However, contrary to the proponents of robustness reasoning, I deny that there is much merit in generating the same observed result using different observational procedures when the relevant differences do not provide such identifiable informational advantages. The convergence of novel, 'independent' observational routes on the same observed result, absent such identifiable informational advantages, might well be completely irrelevant in the assessment of the reliability of these routes. Consider, for example, the independent convergence of pre-Copernican astronomers on the observed result that the earth is stationary. Pre-Copernicans arrived at this observation whenever they stood outside on a windless night and noticed the starry cosmos slowly cycling around the earth. Moreover, they arrived at this observation in a multitude of independent physical circumstances—at different places on the earth, during different seasons, in locales of differing topographies and so on. That is, the observed result—'the earth is stationary and the cosmos revolves around it'—was often and decidedly robustly generated, and the proponent of robustness reasoning is compelled to recognize this result as having a distinct epistemic authority. For me, such a conclusion exhibits the ultimate irrelevance of robustness reasoning as a feature of scientific, observational methodology. There are many ways one might usefully assess the truth of the proposed observed result—here, using a telescope is a particularly worthwhile option (viz., Galileo). Generating the same observed result using a different observational procedure 'just for the sake of it'—just for the sake of using a different, independent procedure—is simply not one of these useful ways.

In phrasing my critique of robustness in these somewhat uncompromising terms, one might be concerned that I have unfairly assessed the epistemic significance of robustness reasoning. Surely a methodological strategy so widely endorsed by philosophers must have some merit (itself an application of robustness reasoning). Yet it is, indeed, this unquestioned orthodoxy harbored by robustness theorists that warrants an unbending critique. Consider, for example, a recent, edited book (Soler et al. 2012) containing many philosophical reflections about, and scientific examples of, robustness reasoning. In the introduction to the book, Lena Soler comments that

> the term 'robustness'...is, today, very often employed within philosophy of science in an intuitive, non-technical and flexible sense that, globally, as acts as a synonym of 'reliable', 'stable', 'effective', 'well-established', 'credible', 'trustworthy', or even 'true'. (3)

One should pause when considering 'robust' as a possible synonym for 'reliable' or 'true', as though one would be speaking nonsense in saying that a robustly generated result is not reliably produced or that it is false. Soler continues by favourably citing the words of Jacob Stegenga in his contribution to the volume:

> Without doubt, the robustness scheme plays an effective and important role in scientific practices. Critics cannot reproach it for being an invention of the philosopher of science. In Stegenga's words, it is 'an exceptionally important notion', 'ubiquitous in science', and a '(trivially) important methodological strategy which scientists frequently use'. (5)

As I see it, however, the wide support for robustness reasoning found in the philosophical literature really is the invention of the philosopher of science. Cynically, it has become a way for philosophers of science to congratulate themselves on finding an abstract method that possesses what Kirshner (2004) calls 'the ring of truth' (265)—the accomplishment of a methodological dream that hearkens back to Descartes' *Discourse on Method*. One should be suspicious of a method that has such broad power. Particularly, one should be suspicious of a method that can be applied in complete ignorance of the details of a scientific case, as is true with robustness reasoning where all one presumably needs to know is that two (or more), minimally reliable observational procedures independently converge on the same result, leaving aside all the technical details of how these procedures arrived at these results.

The derivative burden of this book is to recast the realism/antirealism debate in a novel way, one that hopefully bypasses the polarization that typically afflicts this debate. From the antirealist Bas van Fraassen (1980), we have the skeptical view that accepting a scientific theory amounts to no more than accepting 'that what the theory says about what is observable (for us) is true' (18). The epistemological emphasis that van Fraassen places on human-centred observational capabilities is entirely appropriate and undeniably central to scientific thinking. This emphasis is reflected in the methodological priority I have attached to naked-eye observation. What is less appropriate is van Fraassen's denial that scientists aim, or should aim, at literally true accounts of those parts of the world that aren't revealed by naked-eye observation. For all its methodological priority, the objects revealed through naked-eye observation lack ontological priority. The scientific image of the world is often at odds with its manifest, observed image: Biologists affirm a world of microbes, but unenhanced observation reveals nothing of the sort; astrophysicists believe in dark matter, but naked-eye observation works only with illuminated objects. When it comes to questions of ontology, scientific authority is the arbiter, and scientists do not shy away from ontological commitments that reach far beyond the purview of what is observable 'for us'.

But science is fallible in its ontological commitments, sometimes drastically so, and armed with the pessimistic meta-induction one is prone to conclude that the realist aim is a foolish one. What this means is that a scientist's ontological commitments must always be somewhat tentative, held in abeyance pending further empirical inquiry—and this is exactly how it should be, given how far science currently is from a completely true, comprehensive understanding of the world. So should the realization that we are fallible plunge us into skepticism? This is the tension that pervades the realist/antirealist debate, with the realist overestimating the quality of our scientific theories and the antirealist overestimating our fallibility. It is this tension that I hope to attenuate with methodological preservationism. Even if it happens that our best scientific theory turns out to be mistaken, it is nevertheless maintained and never denied that naked-eye observation is a conduit to learning about the nature of physical reality. Even for van Fraassen (1980), though a theory is (only) empirically adequate, at least 'what it says about observable things and events in this world is true' (12). Here we need to remind van Fraassen and other empiricists that the observable world is not quite so open to unproblematic inspection: Once more, we don't see microbes and see only luminous matter. But the methodological (not the ontological) priority of naked-eye observation is fundamental, and running a close second is a series of relatively uncontroversial, preserved extensions of naked-eye observation, some of which are reason based (e.g., targeted testing and calibration) and others that are technological (e.g., telescopes and microscopes). To these preserved methods scientists always return in their empirical investigations, and the objects they reveal possess for scientists (and for us) prima facie reality. The objects revealed in unenhanced observation are classic in this respect: All manner of smells, tastes, sights and sounds are routinely assumed to have an external reality, and though a sophisticated scientific attitude places their reality solely in the mind, their apparent external reality stubbornly and incessantly intrudes on us. A similar process occurs with preserved extensions of unenhanced observation. It is through such extensions that leads DAMA (through a model-independent process) to affirm the reality of WIMPs, Perrin (on the basis of his vertical emulsion experiments) to defend the reality of atoms, Clowe et al. (in combining light and x-ray telescopy) to suggest the reality of dark matter and HZT (in examining the light curves of very distant supernovae) to exclude the effects of cosmic evolution and extinction in inferring the existence of dark energy. The credibility of these enhanced observational procedures is powerful enough to support realist attitudes in the minds of the participant scientists and their audiences—though of course no one asserts that their ontological announcements are perennially unassailable.

The defense of scientific realism then comes to this. We start with an unchallenged (prima facie) realism about the objects of naked-eye observation (or, more generically, unenhanced observation): It is a realism that no one denies on pain of insanity. We then note the thinly challenged realism afforded the objects of modest, preserved extensions of naked-eye observation. Classic such extensions include the cases of telescopy and microscopy, but we could add the use of such devices as thermometers, weigh scales, rulers, eye glasses and the like—all are technological interventions that to varying degrees are calibrated and target tested by naked-eye observation. Finally we arrive at objects revealed by less authoritative, more conjectural observational procedures, procedures whose lineage shows far less preservation. Such objects (of course, depending on the case) have a less sure ontological status for us—but that could change if the procedures by which they are revealed are shown to be reliable (here calibration and targeted testing play a role, as do a variety of discipline-specific measures). It is accordingly along these lines that we rebut antirealism, for the scope of what is considered real assuredly goes beyond what is observed through unenhanced observation. Moreover, we go beyond antirealism in allowing that even the objects of naked-eye observation could be shown, with scientific progress, to be illusory. Yet this does not lead us to wholesale skepticism. Despite our considered judgments about the fallibility of our unenhanced observational capacities, our irrevocable attachment to the reality of objects revealed through naked-eye observation persists, an attachment buoyed by a pessimistic induction that reveals the epistemic irresponsibility of dogmatic, scientific theorizing.

APPENDIX 1

Proof of (1a), Chapter 1

$$\frac{P(h/e_1 \,\&\, e_2 \,\&\, e_3 \,\&\ldots\&\, e'_j)}{P(h/e_1 \,\&\, e_2 \,\&\, e_3 \,\&\ldots\&\, e_i)}$$

$$= \frac{P(h)\,P(e_1 \,\&\, e_2 \,\&\, e_3 \,\&\ldots\&\, e'_j\,/\,h)}{P(e_1 \,\&\, e_2 \,\&\, e_3 \,\&\ldots\&\, e'_j)} \times \frac{P(e_1 \,\&\, e_2 \,\&\, e_3 \,\&\ldots e_i)}{P(h)\,P(e_1 \,\&\, e_2 \,\&\, e_3 \,\&\ldots\&\, e_i\,/\,h)}$$

$$= \frac{P(e_1 \,\&\, e_2 \,\&\, e_3 \,\&\ldots\&\, e_i)}{P(e_1 \,\&\, e_2 \,\&\, e_3 \,\&\ldots\&\, e'_j)}$$

(NB: $P(e_1 \,\&\, e_2 \,\&\, e_3 \,\&\ldots\&\, e'_j/h) = 1$ and $P(e_1 \,\&\, e_2 \,\&\, e_3 \,\&\ldots\&\, e_i/h) = 1$)

$$= \frac{P(e_1 \,\&\, e_2 \,\&\, e_3 \,\&\ldots\&\, e_m)\,P(e_i\,/\,e_1 \,\&\, e_2 \,\&\, e_3 \,\&\ldots\&\, e_m)}{P(e_1 \,\&\, e_2 \,\&\, e_3 \,\&\ldots\&\, e_m)\,P(e'_j/\,e_1 \,\&\, e_2 \,\&\, e_3 \,\&\ldots\&\, e_m)}$$

$$= \frac{P(e_i\,/\,e_1 \,\&\, e_2 \,\&\, e_3 \,\&\ldots\&\, e_m)}{P(e'_j/\,e_1 \,\&\, e_2 \,\&\, e_3 \,\&\ldots\&\, e_m)}$$

APPENDIX 2

Proofs of (1b) and (1c), Chapter 1

Proof of (1b)

$P(h/e) > P(h/e')$

iff $[P(h/e) \ / \ P(h/e')] > 1$

iff $[P(e/h)\,P(e') \ / \ P(e'/h)\,P(e)] > 1$

iff $P(e/h) \ / \ P(e) > P(e'/h) \ / \ P(e')$

iff $P(e/h) / (P(h)P(e/h) + P(-h)P(e/-h)) > P(e'/h) / (P(h)P(e'/h) + P(-h)P(e'/-h))$

iff $P(e/h) \ / \ P(e/-h) > P(e'/h) \ / \ P(e'/-h)$

Proof of (1c)

$P(h/e_1 \,\&\, e_2 \,\&\, e_3 \,\&\ldots\&\, e_m \,\&\, e_{m+1}) > P(h/e_1 \,\&\, e_2 \,\&\, e_3 \,\&\ldots\&\, e_m \,\&\, e'_j)$

iff $P(e_1 \,\&\, e_2 \,\&\, e_3 \,\&\ldots\&\, e_m \,\&\, e_{m+1}/\, h) \ / \ P(e_1 \,\&\, e_2 \,\&\, e_3 \,\&\ldots\&\, e_m \,\&\, e_{m+1})$
$> P(e_1 \,\&\, e_2 \,\&\, e_3 \,\&\ldots\&\, e_m \,\&\, e'_j/h) \ / \ P(e_1 \,\&\, e_2 \,\&\, e_3 \,\&\ldots\&\, e_m \,\&\, e'_j)$

iff $P(e_1 \,\&\, e_2 \,\&\, e_3 \,\&\ldots\&\, e_m \,\&\, e_{m+1}/h)\ [P(h)\,P(e_1 \,\&\, e_2 \,\&\, e_3 \,\&\ldots\&\, e_m \,\&\, e'_j/h) +$
$P(-h)\,P(e_1 \,\&\, e_2 \,\&\, e_3 \,\&\ldots\&\, e_m \,\&\, e'_j/-h)]$
$> P(e_1 \,\&\, e_2 \,\&\, e_3 \,\&\ldots\&\, e_m \,\&\, e'_j/h)\ [P(h)\,P(e_1 \,\&\, e_2 \,\&\, e_3 \,\&\ldots\&\, e_m \,\&\, e_{m+1}/h)$
$+ P(-h)\,P(e_1 \,\&\, e_2 \,\&\, e_3 \,\&\ldots\&\, e_m \,\&\, e_{m+1}/-h)]$

iff $P(e_1 \,\&\, e_2 \,\&\, e_3 \,\&\ldots\&\, e_m \,\&\, e_{m+1}/h) \ / \ P(e_1 \,\&\, e_2 \,\&\, e_3 \,\&\ldots\&\, e_m \,\&\, e_{m+1}/-h)$
$> P(e_1 \,\&\, e_2 \,\&\, e_3 \,\&\ldots\&\, e_m \,\&\, e'_j\ /h) \ / \ P(e_1 \,\&\, e_2 \,\&\, e_3 \,\&\ldots\&\, e_m \,\&\, e'_j/-h)$

iff $P(e_{m+1}/\, h \,\&\, e_1 \,\&\, e_2 \,\&\, e_3 \,\&\ldots\&\, e_m) \ / \ P(e_{m+1}/-h \,\&\, e_1 \,\&\, e_2 \,\&\, e_3 \,\&\ldots\&\, e_m)$
$> P(e'_j/h \,\&\, e_1 \,\&\, e_2 \,\&\, e_3 \,\&\ldots\&\, e_m) \ / \ P(e'_j/-h \,\&\, e_1 \,\&\, e_2 \,\&\, e_3 \,\&\ldots\&\, e_m)$

APPENDIX 3

Proof of (5a), Chapter 1

$$P(H/R) > P(H/R')$$

$$\text{iff } P(R/H) \,/\, P(R/-H) > P(R'/H) \,/\, P(R'/-H)$$

$$\text{iff } [P(E/H)P(R/E) + P(-E/H)P(R/-E)] \,/\, [P(E/-H)P(R/E) + P(-E/-H)P(R/-E)]$$

$$> [P(E/H)P(R'/E) + P(-E/H)P(R'/-E)] \,/\, [P(E/-H)P(R'/E) + P(-E/-H)P(R'/-E)]$$

$$\text{iff } [P(E/H)P(R/E)P(E/-H)P(R'/E) + P(-E/H)P(R/-E)P(E/-H)P(R'/E) +$$

$$P(E/H)P(R/E)P(-E/-H)P(R'/-E) + P(-E/H)P(R/-E)P(-E/-H)P(R'/-E)]$$

$$> [P(E/H)P(R'/E)P(E/-H)P(R/E) + P(E/-H)P(R/E)P(-E/H)P(R'/-E) +$$

$$P(-E/-H)P(R/-E)P(E/H)P(R'/E) + P(-E/-H)P(R/-E)P(-E/H)P(R'/-E)]$$

$$\text{iff } [P(-E/H)P(R/-E)P(E/-H)P(R'/E) + P(E/H)P(R/E)P(-E/-H)P(R'/-E)]$$

$$> [P(-E/-H)P(R/-E)P(E/H)P(R'/E) + P(E/-H)P(R/E)P(-E/H)P(R'/-E)]$$

$$\text{iff } [P(E/H)P(R/E)P(-E/-H)P(R'/-E) - P(E/-H)P(R/E)P(-E/H)P(R'/-E)]$$

$$> [P(-E/-H)P(R/-E)P(E/H)P(R'/E) - P(-E/H)P(R/-E)P(E/-H)P(R'/E)]$$

$$\text{iff } P(R/E)P(R'/-E)\,[P(E/H)P(-E/-H) - P(E/-H)P(-E/H)]$$

$$> P(R/-E)P(R'/E)\,[P(E/H)P(-E/-H) - P(E/-H)P(-E/H)]$$

$$\text{iff } P(R/E)P(R'/-E) > P(R/-E)P(R'/E)$$

$$\text{iff } P(R/E) \,/\, P(R/-E) > P(R'/E) \,/\, P(R'/-E)$$

APPENDIX 4

Summary of Microbiological Experiments Investigating Mesosomes, 1969–1985, Chapter 2 (adapted from Hudson 1999)

Reference	Preparation	Mesosomes observed?
Remsen (1968)	Freeze-etching, no prep	Yes
Nanninga (1968)	Freeze-etching, glycerol cryoprotection (with or without sucrose)	Yes
Nanninga (1968)	GA (prefix), OsO_4 (fix) freeze-etching or thin section	Yes
Silva (1971)	Thin section, OsO_4 (fix), OsO_4 (with or without calcium, prefix)	Yes
Silva (1971)	Thin section, no OsO_4 (prefix)	Yes

(*Continued*)

(*Continued*)

Reference	Preparation	Mesosomes observed?
Nanninga (1971)	Freeze-fracture, no prep	No
Fooke-Achterrath et al. (1974)	Variety of preparations at 4°C and 37°C	Yes
Higgins and Daneo-Moore (1974)	Freeze-fracture, glycerol cryoprotection or not, GA or OsO_4 (fix)	Yes
Higgins and Daneo-Moore (1974)	Thin section, GA or.1% OsO_4 (prefix), OsO_4 (fix)	Yes
Higgins and Daneo-Moore (1974)	Freeze-fracture, no prep	No
Higgins et al. (1976)	Freeze-fracture, no prep	No
Higgins et al. (1976)	Freeze-fracture, GA (fix)	Yes
Silva et al. (1976)	Thin section, variety of OsO_4 concentrations (prefix or fix)	Yes
Silva et al. (1976)	Thin section, OsO_4, GA then UA (fix)	Yes
Silva et al. (1976)	Thin section, UA as first fixative	No
Silva et al. (1976) and many others	Thin section, unusual treatments (e.g., anaesthetics, antibiotics, etc.)	Yes
Ebersold et al. (1981)	Freeze-fracture, no prep	No

(*Continued*)

(*Continued*)

Reference	Preparation	Mesosomes observed?
Ebersold et al. (1981)	Freeze-substitution, GA, UA and OsO_4 (fix)	No
Dubochet et al. (1983)	Frozen-hydration, no prep	No
Dubochet et al. (1983)	Frozen-hydration, OsO_4 (fix)	Yes
Hobot et al. (1985)	Freeze-substitution, OsO_4 (fix)	No
Ebersold et al. (1981)	Thin-section, using GA and OsO_4 (fix)	Yes

No prep = no OsO_4, GA or UA fixation or prefixation and no cryoprotection (other preparative measures were used); prefix = used at the prefixation stage; fix = used at the fixation stage. See text for further details.

BIBLIOGRAPHY

Abrams, D., et al. (2002), 'Exclusion Limits on the WIMP-Nucleon Cross Section from the Cryogenic Dark Matter Search', *Physical Review* D, 66: 122003.

Abusaidi, R., et al. (2000), 'Exclusion Limits on the WIMP-Nucleon Cross Section from the Cryogenic Dark Matter Search', *Physical Review Letters*, 84: 5699–5703.

Achinstein, P. (2003), *The Book of Evidence*. Oxford: Oxford University Press.

Ahmed, B., et al. (2003), 'The NAIAD Experiment for WIMP Searches at Boulby Mine and Recent Results', *Astroparticle Physics*, 19: 691–702.

Akerib, D., et al. (2004), 'First Results from the Cryogenic Dark Matter Search in the Soudan Underground Lab.' http://arxiv.org/abs/arXiv:astro-ph/0405033, accessed 12 May 2011.

Akerib, D., et al. (2005), 'Exclusion Limits on the WIMP-Nucleon Cross Section from the First Run of the Cryogenic Dark Matter Search in the Soudan Underground Laboratory', *Physical Review* D, 72: 052009.

Alner, G. J., et al. (2005), 'Limits on WIMP Cross-Sections from the NAIAD Experiment at the Boulby Underground Laboratory', *Physics Letters* B, 616: 17–24.

Benoit, A., et al. (2001), 'First Results of the EDELWEISS WIMP Search Using a 320 g Heat-and-Ionization Ge Detector', *Physics Letters* B, 513: 15–22.

Benoit, A., et al. (2002), 'Improved exclusion limits from the EDELWEISS WIMP search', *Physics Letters* B, 545: 43–49.

Bernabei, R., et al. (1998), 'Searching for WIMPs by the Annual Modulation Signature', *Physics Letters* B, 424: 195–201.

Bernabei, R., et al. (1999), 'On a Further Search for a Yearly Modulation of the Rate in Particle Dark Matter Direct Search', *Physics Letters* B, 450: 448–455.

Bernabei, R., et al. (2003), 'Dark Matter Search', *Rivista del Nuovo Cimento* 26: 1–73. http://particleastro.brown.edu/papers/dama0307403astro-ph.pdf, accessed 12 May 2011.

Bernabei, R., et al. (2006), 'Investigating Pseudoscalar and Scalar Dark Matter', *International Journal of Modern Physics* A, 21: 1445–1469.

Bernabei, R., et al. (2008), 'First Results from DAMA/Libra and the Combined Results with DAMA/NaI', *The European Physical Journal* C, 56: 333–355.

Bernadette, B., and I. Stengers (1996), *A History of Chemistry,* translated by D. van Dam. Cambridge, MA: Harvard University Press.

Bovens, L., and S. Hartmann (2003), *Bayesian Epistemology*. Oxford: Oxford University Press.

Calcott, B. (2011), 'Wimsatt and the Robustness Family: Review of Wimsatt's Re-Engineering Philosophy for Limited Beings', *Biology and Philosophy,* 26: 281–293.

Caldwell, R., and M. Kamionkowski (2009), 'Dark Matter and Dark Energy', *Nature,* 458: 587–589.

Campbell, D. T., and D. W. Fiske (1959), 'Convergent and Discriminant Validation by the Multitrait-Multimethod Matrix', *Psychological Bulletin,* 56: 81–105.

Carrier, M. (1989), 'Circles Without Circularity', in J. R. Brown and J. Mittelstrass (eds.), *An Intimate Relation*. Dordrecht, The Netherlands: Reidel, 405–428.

Cartwright, N. (1983), *How the Laws of Physics Lie*. Oxford: Oxford University Press.

Cartwright, N. (1991), 'Replicability, Reproducibility, and Robustness—Comments on Harry Collins', *History of Political Economy,* 23: 143–155.

Chang, H. (2003), 'Preservative Realism and Its Discontents: Revisiting Caloric', *Philosophy of Science,* 70: 902–912.

Chapman G., and J. Hillier (1953), 'Electron Microscopy of Ultrathin Sections of Bacteria', *Journal of Bacteriology,* 66: 362–373.

Clowe, D., A. H. Gonzalez, and M. Markevitch (2004), 'Weak Lensing Mass Reconstruction of the Interacting Cluster 1E0657-556: Direct Evidence for the Existence of Dark Matter', *The Astrophysical Journal,* 604: 596–603.

Clowe, D., M. Bradac, A. H. Gonzalez, M. Markevitch, S. W. Randall, C. Jones, and D. Zaritsky (2006), 'A Direct Empirical Proof of the Existence of Dark Matter', *The Astrophysical Journal,* 648, L109–L113.

Clowe, D., S. W. Randall, and M. Markevitch (2006), 'Catching a Bullet: Direct Evidence for the Existence of Dark Matter', arXiv:astro-ph/0611496v1.

Cosmides, L,. and J. Tooby (1994), 'Origins of Domain Specificity: The Evolution of Functional Organization', in L. A. Hirschfeld and S. A. Gelman (eds.), *Mapping the Mind: Domain Specificity in Cognition and Culture*. Cambridge, MA: Cambridge University Press, 85–116.

Culp, S. (1994), 'Defending Robustness: The Bacterial Mesosome as a Test Case', in David Hull, Micky Forbes, and Richard M. Burian (eds.), *Proceedings of the Biennial Meeting of the Philosophy of Science Association 1994,* Vol. 1. Dordrecht, The Netherlands: Reidel, 46–57.

Culp. S. (1995), 'Objectivity in Experimental Inquiry: Breaking Data-Technique Circles', *Philosophy of Science,* 62: 430–450.

Darwin, C. (1956), 'The Rutherford Memorial Lecture, 1956. The Discovery of Atomic Number', *Proceedings of the Royal Society of London A*, 236: 285–296.

Di Stefano, P., et al. (2001), 'Background Discrimination Capabilities of a Heat and Ionization Germanium Cryogenic Detector', *Astroparticle Physics*, 14: 329–337.

Dodelson, S., and M. Liguori (2006), 'Can Cosmic Structure Form Without Dark Matter?' *Physical Review Letters*, 97: 231301.

Doppelt, G. (2007), 'Reconstructing Scientific Realism to Rebut the Pessimistic Meta- Induction', *Philosophy of Science*, 74: 96–118.

Dretske, F. (1969), *Seeing and Knowing*. Chicago: University of Chicago Press.

Dubochet, J., A. W. McDowell, B. Menge, E. Schmid, and K. G. Lickfeld (1983), 'Electron Microscopy of Frozen-Hydrated Bacteria', *Journal of Bacteriology*, 155: 381–390.

Ebersold, H. R., J. Cordier, and P. Liuthy (1981), 'Bacterial Mesosomes: Method Dependent Artifacts', *Archives of Microbiology*, 130: 19–22.

Einstein, A. (1905), 'On the Movement of Small Particles Suspended in a Stationary Liquid Demanded by the Molecular-Kinetic Theory of Heat', translated by A. D. Cowper, in R. Fürth (ed.), *Investigations on the Theory of Brownian Movement*. New York: Dover, 1–18.

Filippenko, A. (2001), 'Einstein's Biggest Blunder? High-Redshift Supernovae and the Accelerating Universe', *Publications of the Astronomical Society of the Pacific*, 113: 1441–1448.

Filippenko, A., and A. Riess (1998), 'Results from the High-*z* Supernova Search Team', *Physics Reports*, 307: 31–44.

Filippini, J. (2005), 'Big Bang Nucleosynthesis. http://cosmology.berkeley.edu/ Education/ CosmologyEssays/BBN.html, accessed 16 May 2009.

Fodor, J. (1983), *The Modularity of Mind*. Cambridge, MA: MIT Press.

Fodor, J. (1984), 'Observation Reconsidered', *Philosophy of Science*, 51: 23–43.

Fooke-Achterrath, M., K. G. Lickfeld, V. M. Reusch, Jr., U. Aebi, U. Tschope, and B. Menge (1974), 'Close-to-Life Preservation of *Staphylococcus aureus* Mesosomes for Transmission Electron Microscopy', *Journal of Ultrastructural Research*, 49: 270–285.

Franklin, A., and C. Howson (1984), 'Why Do Scientists Prefer to Vary their Experiments?' *Studies in History and Philosophy of Science*, 15: 51–62.

Friedrich, C., D. Moyles, T. J. Beveridge, and R. E. W. Hancock (2000), 'Antibacterial Action of Structurally Diverse Cationic Peptides on Gram-Positive Bacteria', *Antimicrobial Agents and Chemotherapy*, 44: 2086–2092.

Galison, P. (2007), *How Experiments End*. Chicago: University of Chicago Press.

Garnavich, P., et al. (1998), 'Constraints on Cosmological Models from Hubble Space Telescope Observations of High-*z* Supernovae', *The Astrophysical Journal*, 493: L53–L57.

Gates, E. (2009), *Einstein's Telescope*. New York: Norton.

Glanz, J. (1997), 'New Light on Fate of the Universe', *Science*, 278: 799–800.

Greenwood, J. (1990), 'Two Dogmas of Neo-Empiricism: The "Theory-Informity" of Observations and the Quine-Duhem Thesis', *Philosophy of Science*, 57: 553–574.

Hacking, I. (1983), *Representing and Intervening* Cambridge, MA: Cambridge University Press.

Higgins, M. L., and L. Daneo-Moore (1974), 'Factors Influencing the Frequency of Mesosomes Observed in Fixed and Unfixed Cells of Staphylococcusfaecalis', *Journal of Cell Biology*, 61: 288–300.

Higgins, M. L., H. C. Tsien, and L. Daneo-Moore (1976), 'Organization of Mesosomes in Fixed and Unfixed Cells', *Journal of Bacteriology*, 127: 1519–1523.

Hobot, J. A., W. Villiger, J. Escaig, M. Maeder, A. Ryter, and E. Kellenberger (1985), 'Shape and Fine Structure of Nucleoids Observed on Sections of Ultrarapidly Frozen and Cryosubstituted Bacteria', *Journal of Bacteriology*, 162: 960–971.

Howson, C., and A. Franklin (1994), 'Bayesian Conditionalization and Probability Kinematics', *British Journal of the Philosophy of Science*, 45: 451–466.

Howson, C., and P. Urbach (2006), *Scientific Reasoning*. Chicago: Open Court.

Kellenberger E., E. A. Ryter, and J. Séchaud (1958), 'Electron Microscope Study of DNA-Containing Plasms: II. Vegetative and Mature Phage DNA as Compared with Normal Bacterial Nucleoids in Different Physiological States', *Journal of Biophysical and Biochemical Cytology*, 4: 671–678.

Kirshner, R. (2004), *The Extravagant Universe*. Princeton, NJ: Princeton University Press.

Kitcher, P. (1993), *The Advancement of Science*. Oxford: Oxford University Press.

Kosso, P. (1989), 'Science and Objectivity', *The Journal of Philosophy*, 86: 245–257.

Kuhn, T. (1996), *The Structure of Scientific Revolutions*, 3rd ed. Chicago: University of Chicago Press.

Landan, H., Y. Nitzan, and Z. Malik (1993), 'The Antibacterial Activity of Haemin Compared with Cobalt, Zinc and Magnesium Protoporphyrin and its Effect on Potassium Loss and Ultrastructure of *Staphylococcus aureus*', *FEMS Microbiology Letters*, 112: 173–178.

Levins, R. (1966), 'The Strategy of Model Building in Population Biology', *American Scientist*, 54: 421–431.

Lewis, C. I. (1946), *An Analysis of Knowledge and Valuation*. Chicago: Open Court.

Locke, J. (1690), *An Essay Concerning Human Understanding*. New York: Dover Publications.

Mayo, D. (1996), *Error and the Growth of Experimental Knowledge*. Chicago: University of Chicago Press.

McMullin, E. (1984), 'A Case for Scientific Realism', in J. Leplin (ed.), *Scientific Realism*. Berkeley: University of California Press, 8–40.

Milgrom, M. (2002), 'Does Dark Matter Really Exist?', *Scientific American*, August: 43–52.

Milgrom, M. (2008), 'Milgrom's Perspective on the Bullet Cluster', *The MOND Pages*, http://www.astro.umd.edu/~ssm/mond/moti_bullet.html, accessed 24 June 2008.

Moffat, J. (2008), *Reinventing Gravity*. Toronto: Thomas Allen.

Mohr, P. J., B. Taylor, and D. Newell (2008), 'CODATA Recommended Values of the Fundamental Physical Constants: 2006', *Reviews of Modern Physics*, 80: 633–730.

Murrell, J. (2001), '*Avogadro* and His Constant', *Helvetica Chimica Acta*, 84: 1314–1327.

Nanninga, N. (1968), 'Structural Features of Mesosomes (Chondroids) of *Bacillus subtilis* after Freeze-Etching', *Journal of Cell Biology*, 39: 251–263.

Nanninga, N. (1971), 'The Mesosome of Bacillus subtilis as Affected by Chemical and Physical Fixation', *Journal of Cell Biology*, 48: 219–224.

Nanninga, N. (1973), 'Freeze-Fracturing of Microorganism: Physical and Chemical Fixation of *Bacillus Subtilis*', in E. Benedetti and P. Favard (eds.), *Freeze-Etching: Techniques and Applications*. Paris: Societé Française de Microscopie Electronique, 151–179.

Nicolson, I. (2007), *Dark Side of the Universe*. Baltimore, MD: Johns Hopkins University Press.

Nye, M. (1972), *Molecular Reality*. London: Macdonald.

Orzack, S., and E. Sober (1993), 'A Critical Assessment of Levins's *The Strategy of Model Building in Population Biology* (1966)', *The Quarterly Review of Biology*, 68: 533–546.

Perlmutter, S., et al. (1997), 'Measurements of the Cosmological Parameters Ω and Λ from the First Seven Supernovae at $z \geq .35$', *The Astrophysical Journal*, 483: 565–581.

Perlmutter, S., et al. (1998), 'Discovery of a Supernova Explosion at Half the Age of the Universe', *Nature*, 391: 51–54.

Perlmutter, S., et al. (1999), 'Measurements of Ω and Λ from 42 High-Redshift Supernovae', *The Astrophysical Journal*, 517: 565–586.

Perlmutter, S. (2003), 'Dark Energy: Recent Observations and Future Prospects', *Philosophical Transactions of the Royal Society of London* A, 361: 2469–2478.

Perrin, J. (1910), *Brownian Movement and Molecular Reality*, translated by F. Soddy. London: Taylor & Francis.

Perrin, J. (1916), *Atoms*, 4th ed., translated by D. Hammick. New York: D. Van Nostrand.

Perrin, J. (1923), *Atoms*, 11th ed., translated by D. Hammick. New York: D. Van Nostrand.

Perrin, J. (1926), 'Discontinuous Structure of Matter', in *Nobel Lectures: Physics 1922-1941*, Nobel Foundation (ed.). New York: Elsevier, 138–164.

Primack, J. (1999), 'Dark Matter and Structure Formation', in A. Dekel and J. Ostriker (eds.), *Formation of Structure in the Universe*. Cambridge, MA: Cambridge University Press, 3–85.

Psillos, S. (1994), 'A Philosophical Study of the Transition from the Caloric Theory of Heat to Thermodynamics: Resisting the Pessimistic Meta-Induction', *Studies in the History and Philosophy of Science*, 25: 159–190.

Psillos, S. (1999), *Scientific Realism: How Science Tracks Truth*. New York: Routledge.

Rasmussen, N. (1993), 'Facts, Artifacts, and Mesosomes: Practicing Epistemology with a Microscope', *Studies in History and Philosophy of Science*, 24: 227–265.

Rasmussen, N. (2001), 'Evolving Scientific Epistemologies and the Artifacts of Empirical Philosophy of Science: A Reply Concerning Mesosomes', *Biology and Philosophy*, 16: 627–652.

Remsen, C. (1968), 'Fine Structure of the Mesosome and Nucleoid in Frozen-Etched *Bacillus subtilis*', *Archiv fur Mikrobiologie*, 61: 40–47.

Riess, A. (2000), 'The Case for an Accelerating Universe from Supernovae', *Publications of the Astronomical Society of the Pacific*, 112: 1284–1299.

Riess, A., et al. (1998), 'Observational Evidence from Supernovae for an Accelerating Universe and a Cosmological Constant', *The Astronomical Journal*, 116: 1009–1038.

Riess, A., et al. (2004), 'Type Ia Supernovae Discoveries at $z \geq 1$ from the *Hubble Space Telescope*: Evidence for Past Deceleration and Constraints on Dark Energy Evolution', *The Astrophysical Journal*, 607: 665–687.

Ryter, A., and E. Kellenberger (1958), 'Etude au Microscope Électronique de Plasmas Contenant de l'Acide Désoxyribonucléique: I. Les Nucléoides des Bactéries en Croissance Active', *Zeitschrift für Naturforschung*, 13: 597–605.

Salmon, W. (1984), *Scientific Explanation and the Causal Structure of the World*. Princeton, NJ: Princeton University Press.

Sanglard, V., et al. (2005), 'Final Results of the EDELWEISS-I Dark Matter Search with Cryogenic Heat-and-ionization Ge Detectors', *Physical Review* D, 71: 122002.

Santhana, R. L., et al. (2007), 'Mesosomes Are a Definite Event in Antibiotic-Treated Staphylococcus aureus ATCC 25923', *Tropical Biomedicine*, 24: 105–109.

Sanyal, D., and D. Greenwood (1993), 'An Electronmicroscope Study of Glycopeptide Antibiotic-Resistant Strains of *Staphylococcus epidermidis*', *Journal of Medical Microbiology*, 39: 204–210.

Shimoda, M., K. Ohki, Y. Shimamoto, and O. Kohashi (1995), 'Morphology of Defensin-Treated Staphylococcus aureus', *Infection and Immunity*, 63: 2886–2891.

Silva, M. T. (1971), 'Changes Induced in the Ultrastructure of the Cytoplasmic and Intracytoplasmic Membranes of Several Gram-positive Bacteria by Variations of OsO_4 Fixation', *Journal of Microscopy*, 93: 227–232.

Silva, M. T., J. C. F. Sousa, J. J. Polonia, M. A. E. Macedo, and A. M. Parente (1976), 'Bacterial Mesosomes: Real Structures or Artifacts?', *Biochimica et Biophysica Acta*, 443: 92–105.

Sneed, J. (1979), *The Logical Structure of Mathematical Physics*, 2d ed. Dordrecht, The Netherlands: Reidel.

Sober, E. (1999), 'Testability', *Proceedings and Addresses of the American Philosophical Association*, 73: 47–76.

Sober, E. (2008), *Evidence and Evolution*. Cambridge, MA: Cambridge University Press.

Soler, L., E. Trizio, T. Nickles, and W. Wimsatt (eds.) (2012), *Characterizing the Robustness of Science*. Dordrecht, The Netherlands: Springer.

Staley, K. (2004), 'Robust Evidence and Secure Evidence Claims', *Philosophy of Science*, 71: 467–488.

Stanford, K. (2003), 'No Refuge for Realism: Selective Confirmation and the History of Science', *Philosophy of Science*, 70: 913–925.

Stanford, P. K. (2006), *Exceeding Our Grasp*. Oxford: Oxford University Press.

Stegenga, J. (2009), 'Robustness, Discordance, and Relevance', *Philosophy of Science*, 76: 650–661.

Thorndike, E. L. (1920), 'A Constant Error on Psychological Rating', *Journal of Applied Psychology*, 4: 25–29.

Van Fraassen, B. (1980), *The Scientific Image*. New York: Clarendon.

Van Fraassen, B. (2009), 'The Perils of Perrin, in the Hands of Philosophers', *Philosophical Studies*, 143: 5–24.

Virgo, S. (1933), 'Loschmidt's Number', *Science Progress*, 27: 634–649.

Votsis, I. (2011), 'The Prospective Stance in Realism', *Philosophy of Science*, 78: 1223–1234.

Wimsatt, W. (1981), 'Robustness, Reliability, and Overdetermination', in Marilynn B. Brewer and Barry E. Collins (eds.), *Scientific Inquiry and the Social Sciences*. San Francisco, CA: Jossey-Bass, 124–163.

Wisniak, J. (2000), 'Amadeo Avogadro: The Man, the Hypothesis, and the Number', *Chemical Educator*, 2000: 263–268.

Woodward, J. (2006), 'Some Varieties of Robustness', *Journal of Economic Methodology*, 13: 219–240.

Woolgar, S. (2000), *Science: The Very Idea*. London: Routledge.

Worrall, J. (2007), 'Miracles and Models: Why Reports of the Death of Structural Realism May Be Exaggerated', *Royal Institute of Philosophy Supplement*, 61: 125–154.

Wylie, A. (1990), 'Varieties of Archeological Evidence: Gender Politics and Science', paper presented at the Eastern Division meeting of the American Philosophical Association, December 1990, Boston.

Yellin, S. (2002), 'Finding an Upper Limit in the Presence of an Unknown Background', *Physical Review* D, 66: 032005.

INDEX

Abstractness, excessive, 187–188
Accelerative expansion of space, xix, 150–159. *See also* Dark energy
Accuracy, representational, 195–198
Achinstein, Peter, 136
Achterrath-Fooke experiments, 61–62
Annual modulation analysis, 84
Annual modulation result, 95, 101, 179, 233
Anomalies, perception of, 42–43
Antirealism. *See* Realism/antirealism debate
Artifacts. *See* Mesosomes
Atomic theory
- assessment of, 130–134
- Brownian motion and, 116–130
- displacement, rotation, and diffusion and, 124–130
- improved methods and, 234
- overview of, xviii, 3–4, 36, 103–104
- Perrin's table and, 104–107
- preservationism and, 216–217
- realism about molecules and, 134–138
- vertical distributions in emulsions and, 116–124
- viscosity of gases and, 104, 107–116

Atoms (Perrin), xviii, 103
Avogadro's number, xviii, 3–4, 108, 217. *See also* Atomic theory

Babylonian theoretical structures, 25, 27, 192
Bacteria. *See* Mesosomes
Bancelin, Jacques, 130
Bayesian formalisms, 8–24
Bekenstein, Jacob, 143
Bensaude-Vincent, Bernadette, 134
B-H approach. *See* Bovens and Hartmann approach
Big Bang nucleosynthesis, 82
Black, Joseph, 206
Black body radiation, 135
Bovens and Hartmann (B-H) approach, 20–24, 183, 187
Brillouin, Léon, 130
Brownian movement
- displacement, rotation, and diffusion and, 124–130
- vertical distributions in emulsions and, 104, 116–124, 131, 137–138

Brownian Movement and Molecular Reality (Perrin), xviii, 103, 105
Bruner-Postman experiment, 42–43
Bullet Cluster, xix, 142, 238. *See also* Dark matter

Calcott, Brett, xxiii
Calibration
- molecular theory and, xviii–xix, 137, 234

Calibration (*Cont.*)
relevance and, 173
robustness and, 32, 121, 139, 173
Stoke's law, emulsions and, 121
Caloric theory of heat, 203–204, 206–207, 211
Campbell, Donald, 33
Carnot's Principle, 116
Carrier, Martin, 47–48
Cartwright, Nancy, xv, xxiii–xxiv, 193–195
CDMS (Cold Dark Matter Search) group, 83–84, 86–87, 88, 89–91
Cepheid variables, 150–151
Chandra X-ray Observatory, 146
Chang, Hasok, xxi, 203
Chapman, George, 55
Clausius equation, 108, 109, 111
Cline, David, 223
Clowe, Douglas, 144–145
Coasting universe, 153–154
COBE (Cosmic Background Explorer), 155–156
Cognitive impenetrability, 40
Cognitive independence, 177
Cognitive progress, 226
Coincidences, preposterous, 189, 191
Collins, Henry, 34, 35
Colloids, 117
Collusion, 22
Competition, 98
Completeness, 226
Concealed independence, 30, 35
Concurrent processes, 90
Consequentialism, 192–193
Consilience, 226
Conspiracy of fine-tuning, 165
Convergence, spurious, 30, 35
Convergent validation, 34, 58, 99–100
Converse robustness, 179–182, 190
Copernican astronomers, 243
Core argument for robustness
defined, xvii, 7–8
epistemic independence and, 53
independence and, 170–174
Corroborating witness, 182–188
Cosmic accident argument, 3–4
Cosmic Background Explorer. *See* COBE
Cosmic dust, 160, 167, 237–238
Cosmic jerk, 164–165
Cosmological constant, 152–153, 154, 155, 218
Creationism, 37
Critical density, 153
Cryoprotection, 60–65, 66, 72–76
Culp, Sylvia, xvii, 53–54, 56–59, 64
Cumulativism, methodological preservationism and, xxi

DAMA (Dark Matter) group, 80, 82–87, 96, 149–150, 233–234
DAMA/LIBRA experiment, 85
DAMA/NaI experiment, 85
Daneo-Moore, Lolita, 63
Dark energy
in composition of universe, 81
independent convergence and, 177–178
overview of, 152–159
preservationism and, 217–218
robustness and, 166–168
systematic errors and, 159–166
targeted testing and, 141–142, 232–233
Dark matter. *See also* Bullet Cluster; WIMPs
arguments on reality of, xix, 142–150
overview of, 81–82
preservationism and, 217–218
targeted testing and, xix, 141–142, 148–149, 232–233
Dark matter cusps, 142
Dark radiation, 114, 133
Darwin, Charles Galton, 115
Data-technique circles, 58
Degree-of-belief framework, 187
Democracy, 207–208
Density, of emulsive granules, 119–120
Deontology, 192–193
Dialectical strategy, 133–134
Dialectrics, 109
Dick, Rainer, 144–145
Diffusion, Brownian motion and, 124–130
Dimness, 159–160
Direct evidence, 147
Discontinuous structure of matter, 103, 131
Discourse on Method (Descartes), 244
Discriminant validation, 31–34, 35, 184

Displacement, Brownian motion and, 124–130
Divide et impera move, 203–204
Doppelt, Gerald, xxi, 226–228, 238–239
Double-blind tests, 44–45
Dubochet, Jacques, 66–67, 69–70
Duclaux, Jacques, 121
Duhem-Quine nature, 48–49
Dust, cosmic, 160, 167, 237–238

Ebersold, Hans Rudolf, 63, 64, 66
Econometrics, 193–194
EDELWEISS group, 83–84, 86–87, 88, 91–93
Einstein, Albert, 125, 126–131, 152–153, 236
Electromagnetism, ethereal theory of, 202, 203–204, 206–207, 211
Electron, charge of, 133
Electron recoils, 91
Empirical adequacy, 226
Emulsions
  displacement, rotation, and diffusion and, 124–130
  vertical distributions in, 104, 116–124, 131
Epistemic independence
  core argument and, 53
  overview of, xv, xvii, 24
  robustness as based on, 36–51
Epistemic observation, 38–39
Essays Concerning Human Understanding (Locke), xvi
Ethereal theory of electromagnetism, 202, 203–204, 206–207, 211
Euclidean theoretical structures, 25–26
Evolution
  dimness and, 160–161, 164–165
  independence of account and, 37
  modularity and, 40–41
Excessive abstractness, 187–188
Exner, Franz, 126
Expansion of universe, xix, 150–159
Experimental processes and procedures, xxiv. *See also* Observational processes and procedures
Extinction, dimness and, 160–163, 164–165

Failure of robustness, 179–182
Feynman, Richard, 25, 27
Filippenko, Alex, 175
Fire, xvi, 196
Fiske, Donald, 33
Fixation methods, 70–71, 76, 213, 231–232, 236
Flatness hypothesis, 155–156
Flat rotation curves, 82
Fodor, Jerry, 39–40
Fooke-Achterrath experiments, 61–62
Football analogy, 113–114
Forensics example, 182–188
Fraud, 176
Freeze fracturing approach, 63, 67–68
Freeze-substitution approach, 68, 69
Frozen-hydration approach, 66–67, 68
Functional forms, 193–194

Galactic clusters, 82, 142–143. *See also* Bullet Cluster
Galactic rotation curves, 142–145
Galaxies, estimating distance of, 151
Gamma radiation, 88
Gases
  vertical distributions in emulsions and, 116–124, 131, 137–138
  viscosity of, 104, 107–116
Generative approach, 46–49
Germanium detectors, 89–90
Glutaraldehyde, 60–62
Glycerol, 72–75
Goldhaber, Gerson, 167
Gamboge, 118, 121–124, 126–128
Gramme molecules, defined, 108
Gravitational lensing, 146, 160
Gravity
  alternate theories on, 142–145, 147–148
  repulsive, 81, 157–158
Guoy, Louis Georges, 116

Hacking, Ian, xiii–xiv, 2–5, 189–191, 199
Haemin, 63
Halley's comet, 181
Halos, 33, 142, 145, 233
Heat, theories of, 203–204, 206–207, 211

Heat and ionization experiments, 88, 89–90, 91
Henri, Victor, 125, 126–127, 236
Higgins, Michael, 63
High-Z Team. *See* HZT
Hillier, James, 55
Hobot, Jan, 67–68, 69
Hockey analogy, 113
Hubble, Edwin, 150–151
Hubble diagram, 151–152
Hubble Space Telescope, 146
HZT (High-Z Team), 152, 153–164, 166–167, 238

ICM. *See* Intra-cluster medium
Impenetrability, cognitive, 40
Improving standards, 205–207, 226–228. *See also* Methodological preservationism
Inconsistencies, pragmatic approaches to robustness and, 26
Independence
  concealed failure of, 30–31, 32
  core argument for robustness and, 170–174
  defining, xiv–xv
  need for, vs. need for robustness, 174–178
Independence of an account, 36–38, 44–49, 58
Independent angles, 188
Indeterminism, mesosomes and, 72–78
Inferential robustness, 27–28, 193–194
Internal coherence, 226
Intra-cluster medium (ICM), 146–147
Intuitive plausibility, 226, 228

Jerk, cosmic, 164–165

K-corrections, 160
Keesom, Willem, 132
Kirshner, Robert, xix, 142, 156, 166–167, 243–244
Kosso, Peter, xvii, 36–37
Kuhn, Thomas, 41–42

Lavoisier, Antoine, 206
Leeds, Steve, 15
Lensing, gravitational, 146, 160
Levins, Richard, xxiii
Lewis, C.I., 20–21
LIBRA experiment, 85
Light curves, 152–153
Locke, John, xvi, 196
Logic, lack of robustness in, 189–195
Loschmidt, Josef, 114–115
Low mass-density universe, 158–159, 160
Luminosity, 143, 146–147, 159–160, 225, 238. *See also* Dark matter

Magnification, 231
Magnitude. *See* "Order of magnitude" robustness argument
Malmquist bias, 160
Mass, calculations of, 120
Mathematics, lack of robustness in, 189–195
Maxwell, J.C., 114, 115
Maxwell's equations, 107–108, 206, 207, 211
Mayo, Deborah, 33
McMullin, Ernan, 221–223
Mean free path, 107–108, 110
Measurement robustness, 194–195
Meditations (Descartes), 231
Mesosomes
  experiments on, 59–65
  indeterminism and, 72–78
  overview of, xvii, 52–55
  preservationism and, 213–215
  Rasmussen and Culp and, 55–59
  reliable process reasoning and, 65–72
  representational accuracy and, 196–197
  targeted testing and, 149
Meta-induction, pessimistic, 218–225
Methodological preservationism, xxi, 202, 226, 240–243
Microbiology. *See* Mesosomes
Microscopy, xxii, 3–5. *See also* Mesosomes
Milgrom, Mordehai, 142–143, 148
Millikan, Robert, 133
Minimal reliability requirement, 18, 22, 57, 106, 174, 199, 200, 230
Miracles. *See* No-miracles arguments
Model-dependent observational research, 80, 84, 87, 88–93, 233

Model-independent observational research, 80–81, 82–87, 233
Modularity of perception, 39–40
Modulation effects, 85–86
Molecular motion, 203
Molecular theory
  assessment of, 130–134
  Brownian motion and, 116–130
  displacement, rotation, and diffusion and, 124–130
  overview of, xviii, 3–4, 36, 103–104
  Perrin's table and, 104–107
  realism about molecules and, 134–138
  vertical distributions in emulsions and, 116–124
  viscosity of gases and, 104, 107–116
Moles, defined, 108
MOND (Modified Newtonian Dynamics) theory, 143–144, 147–148
Morality, 192–193, 201
Mossotti's theory of dialectrics, 109
Müller-Lyer Illusion, 40, 41
Multiple derivations, 192
Multiple scatterings, 90
Muons, 83, 85, 90, 91
Muon veto, 90, 94
Murrell, John, 114

NaI (T1)(Thallium-activated sodium iodide), 83
NAIAD trial, 88
Naked-eye observation, xxiv, 231, 234–235, 237–239, 242–247
Nanninga, Nanne, 60, 63, 72–77
Negative results, 57
Neutralinos. *See* WIMPs
New induction, 203
Newtonian mechanics, 202
Newton's second law, 47–48
Nicolson, Iain, 142–144, 156–157
Nobel Prize, 121, 130, 141
No-miracles arguments
  for realism, 202–204
  for robustness, 1, 2–8
Nonepistemic observation, 38–39, 43, 56
Nuclear recoils, 88–89, 90, 91–92

Objective probability, 13, 23
Objectivity, main threat to, xv
Observation, epistemic vs. nonepistemic, 38–39, 43
Observational processes and procedures, defined, xxiv
Observational robustness, xxi, 228–229
Oddie, Graham, 15–16
Oddie-Leeds (OL) formalism, 15–18
OL formalism. *See* Oddie-Leeds (OL) formalism
Optimum interval method, 92–93
"Order of magnitude" robustness argument, 111–114, 115–116, 132
Organelles. *See* Mesosomes
Orzack, Steven, 209–210
Osmium tetroxide, 60–62, 65–66, 69–70, 75–77, 213–215

Perception, 39–40, 42–43
Peripheral bodies. *See* Mesosomes
Perlmutter, Saul, 141, 156, 159, 200
Perrin, Jean, xxiv, 36, 216–217. *See also* Molecular theory
Perrin's table, 104–107
Pessimistic meta-induction, 218–225
Physical independence, overview of, xv
Platonic dialogues, 207
Plausibility, intuitive, 226, 228
Polymerization, 60
Pragmatic approaches to robustness, 25–36
Pragmatic reliability, 26–27
P(REL). *See* Probability (that a witness is reliable)
Preposterous coincidences, 189, 191
Presentism, 210
Preservationism. *See also* Methodological preservationism; Theoretical preservationism
  atoms and, 216–217
  dark matter, dark energy and, 217–218
  defense of realism using, xx–xxi, 204
  mesosomes and, 213–215
  pessimistic meta-induction and, 218–225
  WIMPs and, 215–216
Pressure, gases, emulsions and, 117–119
Probabilistic approaches to robustness, 8–24

Probability (that a witness is reliable) (P(REL)), 21
Psychology, 175–176
Pulse shape discrimination, 83, 88–89, 90
Pylyshyn, Zenon, 40

Radioactivity, 133
Radon gas, 85–86
Rasmussen, Nicolas, 54–59, 71–78
Rationality, Rasmussen on, 72
Rayleigh (Lord), 104, 132–133
Realism, structural, 204–206
Realism/antirealism debate
  arguments against theoretical preservationism and, 208–218
  arguments for theoretical preservationism and, 204–208
  methodological preservationism and, 226–243
  no-miracles argument for realism and, 202–204
  overview of, 201–202, 245–246
  pessimistic meta-induction, preservationism and, 218–225
Received model, 153–154
Red blood cell example, 3–5
Redshifts, 151–152, 153–154, 238
Redundancy, 26
Relevance, independence and the core argument and, 172
Reliability. *See also* Minimal reliability requirement
  mesosomes and, 57–58
  modularity and, 40
  overview of, 5–8
  pragmatic approaches to robustness and, 26–27
  probabilistic approaches to robustness and, 10–13, 23
Reliable process reasoning
  expansion of universe and, 162–163
  importance of, 229–230
  mesosome example and, xvii, 54, 65–72
  molecular theory example and, 127
  WIMPs example and, xviii, 97–102
Remsen, Charles, 60
Replicability, 180–182
Representing and Intervening (Hacking), xiii–xiv
Repulsive gravity, 81, 157–158
Riess, Adam, 141, 163, 200
Ring of truth, 156, 174, 180, 200, 244
R-K fixation. *See* Ryter-Kellenberger fixation
Robust detection, definition of robustness and, xxiii
Robustness
  corroborating witnesses and, 182–188
  definitions of, xxii–xxiii
  epistemic independence approaches to, 36–51
  failure to ground representational accuracy of, 195–198
  independence and the core argument and, 170–174
  lack of in mathematics and logic, 189–195
  need for, vs. need for independence, 174–178
  no-miracles argument for, 1, 2–8
  overview of arguments for and against, 1–2, 51
  pragmatic approaches to, 25–36
  probabilistic approaches to, 8–24
  resistance to converse of, 179–182
  sociological dimension of, 198–200
Robust theorem, xxiii
Rotation of Brownian particles, 129–130
Rotation curves, 82, 84, 142–145
Rumford (Count), 196
Ryter-Kellenberger (R-K) fixation, 70–71, 76, 213, 231–232, 236

Salmon, Wesley, xv
Schmidt, Brian, 141, 152, 200
Scientific realism. *See* Realism/antirealism debate
SCP (Supernova Cosmology Project), 152, 153–162, 166–167
Sectioning process, 60
Seddig, Max, 126
Selection bias, 160
Silicon detectors, 89–90
Silva, Marcus, 61, 63, 65–66
Simplicity, 226, 228
Skin, cellular structure of, 235–236

Sky, blueness of, 104, 132–133
Smoluchowski, Marian, 132, 185
Sneed, Joseph, 47
SN Ia. *See* Supernovae type Ia
Sober, Elliott, 209–210
Sober approach, 18–20, 22–24
Sociological dimension of robustness, 175–176, 198–200
Soler, Lena, 244
Spiral galaxies, 81–82, 143–144
Spurious convergence, 30, 35
Staley, Kent, 24, 28–36
Standards, 240–241
Standards, improving, 205–207, 226–228. *See also* Methodological preservationism
Standards of explanatory and predictive success, 226, 228
Standards preservationism. *See* Methodological preservationism
Stanford, Kyle, xxi, 202–203, 209
Stegenga, Jacob, 243
Stengershas, Isabelle, 134
Stoke's law, 120, 121, 125, 128–130
Structural realism, 204–206
The Structure of Scientific Revolutions (Kuhn), 41–42
Subjective probability, 23–24
Summing example, 189–192
Suntzeff, Nick, 152
Supernova Cosmology Project. *See* SCP
Supernovae type Ia, 141, 151–159
Surface electron events, 90–91
Svedberg, Theodor, 114, 126–127
Systematic errors, dark energy and, 159–166

Targeted testing
  dark energy and, 141–142, 232–233
  dark matter and, xix, 141–142, 148–149, 232–233
  mesosomes and, 149
  observational claims and, 28
  overview of, 141–142
  relevance and, 173
  reliability and, 185, 186, 188
  underdetermination problems and, 197
  WIMP detection and, 149–150
Teicoplanin, 63–64
Telescopy, xxii, 146, 164, 229–230, 232, 243
TeVeS (Tensor-Vector-Scalar field theory), 143, 147–148
Thallium-activated sodium iodide (NaI (T1)), 83
Theoretical preservationism
  arguments against, 208–218
  arguments for, 204–208
  overview of, 203
Thermodynamics, Second Law of, 116
Thermometer example, 15, 172, 195, 247
Triangulation, xiii, 170
Truth, ring of, 156, 174, 180, 200, 244
Tully-Fisher relationship, 143

UA. *See* Uranyl acetate
UKDM (United Kingdom Dark Matter) group, 86–87, 88–89, 95–96
Uncertainty, Perrin's calculations and, 110–111, 124
Underdetermination argument, 197, 202–203
Unenhanced observation, 231
Unification, 226, 228
Universe
  expansion of, xix, 150–151, 153
  low mass-density, 158–159
Uranyl acetate (UA), 63, 65–66, 69

Validation, discriminant, 31–34, 35, 184
Vancomycin, 63
Van der Waals equation, 110–111
Van Fraassen, Bas, 135–136, 245–246
Van't Hoff's law, 117
Viscosity, of gases, 104, 107–116

Whiggism, 210
Wilkinson Microwave Anisotropy Project. *See* WMAP
WIMP halo, 84
WIMPs (weakly interacting massive particles)
  DAMA model-independent approach to detecting, 82–87
  dark matter and, 81–82

WIMPs (weakly interacting massive particles) (*Cont.*)
  historical argument against robustness and, 93–97
  improved methods and, 232, 233
  model-dependent approaches to detecting, 88–93
  overview of, xviii, 79–81
  preservationism and, 215–216
  reliable process reasoning and, xviii, 97–102
  targeted testing and, 149–150
WIMP wind, 84–85
Wimsatt, William, xvii, 24, 29
Witness, corroborating, 182–188
WMAP (Wilkinson Microwave Anisotropy Project), 155–156
Woodward, Jim, 193–195
Woolgar, Steve, 170–171
Worrall, John, 204–206
Yellin method, 92–93
Zwicky, Fritz, 217–218